RANEY NICKEL-ASSISTED SYNTHESIS OF HETEROCYCLES

RANEY NICKEL-ASSISTED SYNTHESIS OF HETEROCYCLES

NAVJEET KAUR
Department of Chemistry at Banasthali Vidyapith, India

Elsevier
Radarweg 29, PO Box 211, 1000 AE Amsterdam, Netherlands
The Boulevard, Langford Lane, Kidlington, Oxford OX5 1GB, United Kingdom
50 Hampshire Street, 5th Floor, Cambridge, MA 02139, United States

Copyright © 2022 Elsevier Inc. All rights reserved.

No part of this publication may be reproduced or transmitted in any form or by any means, electronic or mechanical, including photocopying, recording, or any information storage and retrieval system, without permission in writing from the publisher. Details on how to seek permission, further information about the Publisher's permissions policies and our arrangements with organizations such as the Copyright Clearance Center and the Copyright Licensing Agency, can be found at our website: www.elsevier.com/permissions.

This book and the individual contributions contained in it are protected under copyright by the Publisher (other than as may be noted herein).

Notices
Knowledge and best practice in this field are constantly changing. As new research and experience broaden our understanding, changes in research methods, professional practices, or medical treatment may become necessary.

Practitioners and researchers must always rely on their own experience and knowledge in evaluating and using any information, methods, compounds, or experiments described herein. In using such information or methods they should be mindful of their own safety and the safety of others, including parties for whom they have a professional responsibility.

To the fullest extent of the law, neither the Publisher nor the authors, contributors, or editors, assume any liability for any injury and/or damage to persons or property as a matter of products liability, negligence or otherwise, or from any use or operation of any methods, products, instructions, or ideas contained in the material herein.

ISBN: 978-0-323-99492-7

For Information on all Elsevier publications visit our website at
https://www.elsevier.com/books-and-journals

Publisher: Candice Janco
Editorial Project Manager: Judith Clarisse Punzalan
Production Project Manager: Paul Prasad Chandramohan
Cover Designer: Mark Rogers

Typeset by Aptara, New Delhi, India

CONTENTS

Preface — vii
About the author — ix
Abbreviations — xi

1 Raney nickel-assisted nitro group reduction for the synthesis of five-membered *N*-heterocycles — 1

1.1 Introduction — 1
1.2 Nitro group reduction involved in the synthesis of pyrroles — 2
1.3 Nitro group reduction involved in the synthesis of fused pyrroles — 10
1.4 Nitro group reduction involved in the synthesis of indoles — 14
1.5 Nitro group reduction involved in the synthesis of fused indoles — 18
1.6 Nitro group reduction involved in the synthesis of pyrazoles — 20
1.7 Nitro group reduction involved in the synthesis of imidazoles and benzimidazoles — 22
1.8 Nitro group reduction involved in the synthesis of triazoles — 33
References — 36

2 Raney nickel-assisted nitro group reduction for the synthesis of *N*-, *O*-, and *S*-heterocycles — 43

2.1 Introduction — 43
2.2 Reduction of NO_2 group for the synthesis of five-membered *O*- and *S*-heterocycles — 43
2.3 Reduction of NO_2 group for the synthesis of six-membered *N*-heterocycles — 45
2.4 Reduction of NO_2 group for the synthesis of six-membered *N,N*-heterocycles — 55
2.5 Reduction of NO_2 group for the synthesis of six-membered *N,N,N*-heterocycles — 62
2.6 Reduction of NO_2 group for the synthesis of six-membered *O*-heterocycles — 62
2.7 Reduction of NO_2 group for the synthesis of seven-membered heterocycles — 66
References — 74

3 Synthesis of heterocycles from cyanide, oxime, and azo compounds using Raney nickel — 81

3.1 Introduction — 81
3.2 Synthesis of heterocycles from cyanide compounds — 82

	3.3 Synthesis of heterocycles from oxime compounds	101
	3.4 Synthesis of heterocycles from azo compounds	107
	References	111

4 Synthesis of heterocycles from oxazoles and oxazines using Raney nickel — 119

	4.1 Introduction	119
	4.2 Synthesis of five-membered *N*-heterocycles from oxazoles	120
	4.3 Synthesis of five-membered *N*-heterocycles from oxazines	124
	4.4 Synthesis of five-membered fused *N*-heterocycles from oxazoles and oxazines	128
	4.5 Synthesis of five-membered *O*-heterocycles from oxazoles and oxazines	141
	4.6 Synthesis of six-membered heterocycles from oxazoles and oxazines	142
	4.7 Synthesis of higher-membered heterocycles from oxazoles and oxazines	152
	References	153

5 Miscellaneous use of Raney nickel for the synthesis of heterocycles — 159

	5.1 Introduction	159
	5.2 Synthesis of five-membered *N*-heterocycles	160
	5.3 Synthesis of five-membered *N*-polyheterocycles	161
	5.4 Synthesis of five-membered fused *N*-heterocycles	162
	5.5 Synthesis of five-membered *N,N*-heterocycles	169
	5.6 Synthesis of five-membered *N,N,N*-heterocycles	170
	5.7 Synthesis of five-membered *O*-heterocycles	171
	5.8 Synthesis of five-membered *O,N*- and *S*-heterocycles	174
	5.9 Synthesis of six-membered *N*-heterocycles	177
	5.10 Synthesis of six-membered *N,N*-heterocycles	195
	5.11 Synthesis of six-membered *O*-heterocycles	197
	5.12 Synthesis of seven-membered heterocycles	200
	References	202

Conclusion	*211*
Index	*213*

Preface

Heterocyclic chemistry is an inexhaustible source of novel compounds as an applied science. The compounds with most diverse chemical, physical, and biological properties are provided by designing a wide range of combinations of carbon, hydrogen, and heteroatoms.

Heterocycles are ubiquitous in organic materials, pharmaceuticals, various functional molecules, and natural products. The *N*-, *O*-, and *S*-heterocycles are being studied for a long time due to their biological properties. The literature is enriched with advanced findings about the formation and pharmacological activities of heterocycles. There are a number of heterocyclic natural products such as alkaloids, antibiotics, cardiac glycosides, and pesticides having a lot of importance for human and animal health. Therefore, natural models have been followed for designing and constructing weed killers, pharmaceuticals, pesticides, rodenticides, and insecticides.

Among heterocyclic compounds, aromatic heterocyclic compounds represent structural motifs found in a great number of biologically active synthetic and natural compounds, agrochemicals, and medicines. Moreover, aromatic heterocyclic compounds are extensively utilized for the formation of dyes and polymeric materials of great importance. In organic synthesis, there are several reports on the use of aromatic heterocyclic compounds as intermediates.

Although a wide range of highly efficient approaches have been described in the past for the formation of aromatic heterocyclic compounds and their derivatives, the development of new procedures is in continuous demand. Particularly, development of novel synthetic methodologies toward heterocyclic compounds, aiming at attaining better levels of molecular complexity and improved functional group compatibilities in a convergent and atom economical fashion from easily available starting compounds under mild reaction conditions, is one of the major research activities in synthetic organic chemistry.

As a result, the ongoing interest for developing new efficient and versatile preparation of heterocycles has always been a thread in the synthetic area. In the last years the creative ideas of multicomponent procedures, domino reactions and sequential reactions, where complex and highly diverse structures are produced in a one-pot manner, have significantly stimulated both industry and academia.

Significant focus has been paid to novel approaches for the formation of heterocyclic compounds. The heterogeneous nickel catalyst is used in asymmetric synthesis. Without doubt, Ra-Ni catalyst has been the most utilized catalyst in this context. Raney nickel has been utilized for a broad range of reactions such as methanation, hydrogenolysis, steam reforming, hydrogenation, reductive amination, and desulfurization, either on an industrial scale or in the laboratory. In this book, I have focused on the utilization of Raney nickel for the synthesis of heterocyclic compounds.

About the author

Dr. Navjeet Kaur, daughter of Sardar Sewak Singh and Sardarni Harmesh Kaur, was born in Punjab, India. After attaining her school education, she received her BSc from Panjab University Chandigarh (Punjab, India) in 2008. In 2010, she completed her MSc in Chemistry from Banasthali Vidyapith. She was awarded with PhD in 2014 by the same university under the supervision of **Prof. D. Kishore**. Being a meritorious student, she won a merit scholarship for her master's and doctorate's degrees. Presently, she is working as an Assistant Professor in Department of Chemistry, Banasthali Vidyapith and has entered into a specialized research career focused on the synthesis of 1,4-benzodiazepine-based heterocyclic compounds (Organic Synthetic and Medicinal Chemistry). **Dr. Kaur** has guided many MSc dissertation students and is currently guiding 5 research scholars as their PhD supervisor. With 11 years of teaching experience, she has published over 160 scientific research papers, review articles, book chapters, and monographs in the field of organic synthesis in national and international reputed journals. She has published five authored books: "Palladium Assisted Synthesis of Heterocycles" and "Metals and Non-metals: Five-Membered *N*-Heterocycle Synthesis" with CRC Press, Taylor & Francis group; "Metal- and Nonmetal-Assisted Synthesis of Six-Membered Heterocycles" and "Raney Nickel-Assisted Synthesis of Heterocycles" with Elsevier; "Lawesson's reagent in heterocycle synthesis" with Springer Nature. Her name has featured consistently four times in the **WORLD RANKING OF TOP 2% SCIENTISTS** in the subject-wise analyses conducted by a team of scientists at **Stanford University**, USA. She secured her place among top 2% scientists of the world in **2017** (single year ranking), **2018** (full career wise ranking), **2019** [full career wise ranking with 424 world rank (04 in India) and single year ranking with 03 world rank (01 in India)], and **2020** (full career wise ranking with 372 world rank and single year ranking with 126 world rank). This story of achieving top ranking in such a short span (5–6 years) of her career was covered by numerous newspapers. She was presented the Prof. G. L. Telesara Award in 2011 by Indian Council of Chemists (Agra, Uttar Pradesh) at Osmania University (Hyderabad), and the Best Paper Presentation Award in National Conference on "Emerging Trends in Chemical and Pharmaceutical Sciences" (Banasthali Vidyapith, Rajasthan). She has been cited as the most productive and distinguished author of Banasthali Vidyapith in "Scientometrics Profile of the Banasthali Vidyapith: A Deemed University of Rajasthan, India" (2020), Library Philosophy and Practice (e-journal). In 2021, Dr. Kaur was chosen as a candidate for membership by Royal Society of Research, London, England, UK with full waiver of application fees. She is serving as an Editor-in-Chief for "Advanced Chemicobiology Research" journal. She has attended about 45 conferences, workshops, seminars, and refresher courses; using these platforms she has disseminated her research work through oral and poster presentations. She has delivered many invited lectures and radio talks. She acted as a reviewer for many papers of different journals. She has served as a member of organizing committee of conferences conducted by her department and acted as a member of national advisory committee in International Conference organized by Akal University, Punjab in collaboration with the Indian Chemical Society. Apart from all this, she has been working as NSS (National Service Scheme) Program Officer since 2016 and member of UBA (Unnat Bharat Abhiyan) since 2018. **Dr. Navjeet** finds interest in Sikh literature and has completed a 2-year Sikh Missionary course from Sikh Missionary College (Ludhiana, Punjab).

Abbreviations

ADME	absorption, distribution, metabolism, and excretion
AIBN	azobis-isobutyronitrile
BBDM	*t*-butoxybis(dimethylamino)methane
BINOL	1,1'-bi-2-naphthol
Boc	*t*-butoxycarbonyl
BOP	benzotriazol-l-yloxytris(dimethylamino)phosphonium hexafluorophosphate
CAN	ceric ammonium nitrate
Cbz	carboxybenzyl
CDC	cross-dehydrogenative coupling
CMC	comprehensive medicinal chemistry
CPA	cyclopiazonic acid
CSA	camphorsulfonic acid
DABCO	1,4-diazabicyclo[2.2.2]octane
DBN	1,5-diazabicyclo[4.3.0]non-5-ene
DBU	1,8-diazabicyclo[5.4.0]undec-7-ene
DCC	*N,N*'-dicyclohexylcarbodiimide
DCE	dichloroethane
DCM	dichloromethane
DCP	2,4-dichloropheoxyacetic acid
DDQ	2,3-dichloro-5,6-dicyanobenzoquinone
DEAD	diethyl azodicarboxylate
DEAE	diethylethanolamine
DIAD	diisopropyl azodicarboxylate
DIBAH	diisobutylaluminium hydride
DIBAL	diisobutylaluminium hydride
DIBAL-H	diisobutylaluminium hydride
DIEA	*N,N*-diisopropylethylamine
DIPEA	*N,N*-diisopropylethylamine
DMA	dimethylaniline/dimethylacetamide
DMAE	dimethylethanolamine
DMAP	4-dimethylaminopyridine
DME	dimethoxyethane
DMF	dimethylformamide
DMF-DMA	dimethylformamide dimethylacetal
DMP	Dess-Martin periodinane/dimethoxypyridine/2,9-dimethyl-1,10-phenanthroline
DMSO	dimethylsulfoxide
Dppf	1,1'-bis(diphenylphosphino)ferrocene
EDAC	1-ethyl-3-(3-dimethylaminopropyl)carbodiimide hydrochloride
EDC	1-ethyl-3-(3-dimethylaminopropyl)carbodiimide
GABA	γ-aminobutyric acid
HATU	hexafluorophosphate azabenzotriazole tetramethyluronium
HMPA	hexamethylphosphoramide

HPLC	high performance liquid chromatography
5-HT1A	5-hydroxytryptamine
IBX	*o*-iodoxybenzoic acid
IPA	isopropyl alcohol
KHMDS	potassium hexamethyldisilazide
KMDS	potassium hexamethyldisilazide
LA	Lewis acid
LDA	lithium diisopropylamide
LHMDS	lithium hexamethyldisilazide
LiHMDS	lithium hexamethyldisilazide
LPL	lipoprotein lipase
MAHTs	malonic half thioesters
MAPh	methylaluminium-bis(2,6-diphenylphenoxide)
MBO	2-methyl-3-buten-2-ol
MOM	methoxymethyl
MVK	methyl vinyl ketone
MW	microwave
NaHMDS	sodium hexamethyldisilazide
NBS	*N*-bromosuccinimide
NCS	*N*-chlorosuccinimide
NMM	*N*-methylmorpholine
NMO	*N*-methylmorpholine-*N*-oxide
NMP	*N*-methylpyrrolidinone
NMR	nuclear magnetic resonance
NOE	nuclear Overhauser effect
OTf	trifluoromethanesulfonate
PBD	pyrrolobenzodiazepine
PDC	pyridinium dichromate
PEG	poly(ethylene glycol)
pHTX	philanthotoxin
PIDA	phenyliodine diacetate
Piv	pivaloyl
PMB	*p*-methoxybenzylamine
PMP	polymethylpentene
PPE	polyphosphoric ester
PPTS	pyridinium *p*-tolylsulfonate
PTSA	*p*-toluenesulfonic acid
PZQ	praziquantel/pyrazino-isoquinoline
QCS	quinolinium camphorsulfonate
RCM	ring-closing metathesis
SAR	structure-activity relationship
SEM	[2-(trimethylsilyl)ethoxy]methyl
S_N2	bimolecular nucleophilic substitution
TBAB	tetrabutylammonium bromide
TBAF	tetrabutylammonium fluoride
TBAI	tetrabutylammonium iodide
TBDMS	*t*-butyldimethylsilyl

TBDPS	*t*-butyldiphenylsilyl
TBS	*t*-butyldimethylsilyl
TBSOTf	*t*-butyldimethylsilyl trifluoromethanesulfonate
TCA	trichloroacetic acid
TCC	(cumyl)cyclohexanol
TDS	dimethylhexylsilyl
TEA	triethylamine
TEAC	tetraethylammonium chloride
TEBAC	triethylbenzylammonium chloride
TEMPO	(2,2,6,6-tetramethylpiperidin-1-yl)oxyl or (2,2,6,6-tetramethylpiperidin-1-yl)oxidanyl
TES	triethylsilyl
TFA	trifluoroacetic acid
TFAA	trifluoroacetic anhydride
THF	tetrahydrofuran
THP	tetrahydropyran
THSβCs	tetrahydrospiro-beta-carbolines
TIPS	triisopropylsilyl
TMEDA	tetramethylethylenediamine
TMG	1,1,3,3-tetramethylguanidine
TMS	trimethylsilyl/tetramethylsilane
TMSCN	trimethylsilyl cyanide
TMSI	trimethylsilyl iodide
TOSMIC	toluenesulfonylmethyl isocyanide
TPAP	tetra-*n*-propylammonium perruthenate
TPP	triphenylphosphine/tetraphenylporphyrin

CHAPTER 1

Raney nickel-assisted nitro group reduction for the synthesis of five-membered *N*-heterocycles

1.1 Introduction

The *N*-containing heterocyclic compounds have attracted attention in the past few decades due to their high therapeutic values [1]. Whether they are synthetic or natural one, they are elaborated as important constituents in biological procedures due to their remarkable biological features [2]. Various *N*-containing heterocyclic compounds particularly, in plant kingdom have made indelible mark as phytochemical medicines like theophylline, quinine, emetine, procaine, ellipticine, codeine, morphine, and papaverine [3]. Besides, the massive distribution of *N*-containing heterocyclic compounds in natural products, they also play a key role in the biochemical procedures in living cells. Most of the enzymes have aromatic heterocyclic compounds as main elements while most of the coenzymes having nonamino acids functionalities are aromatic *N*-containing heterocyclic compounds and some important vitamins are formed on aromatic heterocyclic framework [4–6]. Additionally, *N*-containing heterocyclic compounds have been often found as an important structural unit in synthetic medicines like antipyrine, chlorpromazine, metronidazole, captopril, chloroquinine, barbituric acid, isoniazid, azidothymidine, and diazepam. Significant focus has been paid to novel approaches for the formation of *N*-containing heterocyclic compounds, which are natural biologically active organic compounds or pharmacophoric fragments [7–9].

Their synthesis needs an incorporation of *N*-containing functionality via C–C and C–N bond formation processes. The NO_2 compounds act as important building blocks in the formation of nitrogen-containing heterocyclic compounds because of their high chemical reactivity [10].

The heterogeneous nickel catalyst is used in an asymmetric synthesis. Heterogeneous catalytic reactions, as compared to their homogeneous

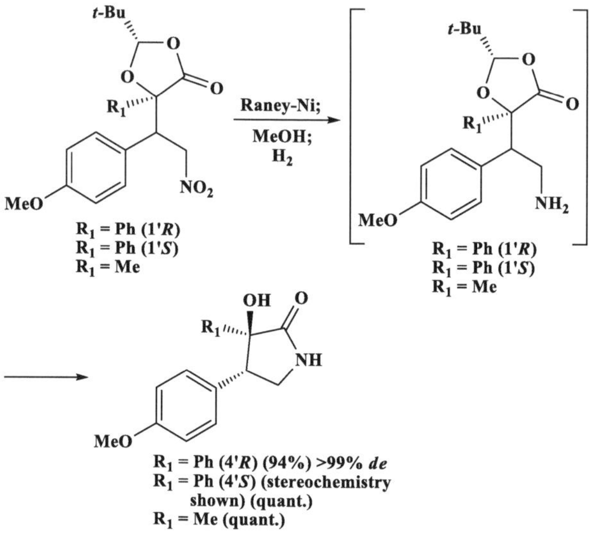

Scheme 1.1

equivalents, have some specific properties because they proceed on a solid surface [11].

1.2 Nitro group reduction involved in the synthesis of pyrroles

Different reaction conditions were applied in an effort to achieve this reaction starting with 10% Pd/C under hydrogen atmosphere in $CH_3COOC_2H_5$, which give rise to no reaction. The conditions were modified to some extent by changing the reaction solvent to 1 M hydrochloric acid and the results for the reaction were inconclusive in this case. Then, $NaBH_4$ was incorporated into reaction with 10% Pd/C in tetrahydrofuran but only starting compound was found. The unreacted compound from this reaction was then recycled and reacted with Raney-Ni in EtOH/tetrahydrofuran solution under nitrogen atmosphere, which was inconclusive again (Scheme 1.1). However, it was decided that the Raney-Ni may not have been very active, therefore; the reaction was carried out in MeOH under hydrogen atmosphere. Simultaneous intramolecular aminolysis of acetal under these conditions provided γ-lactams via cyclization of amine, which led to elimination of pivalaldehyde (Scheme 1.2) [12].

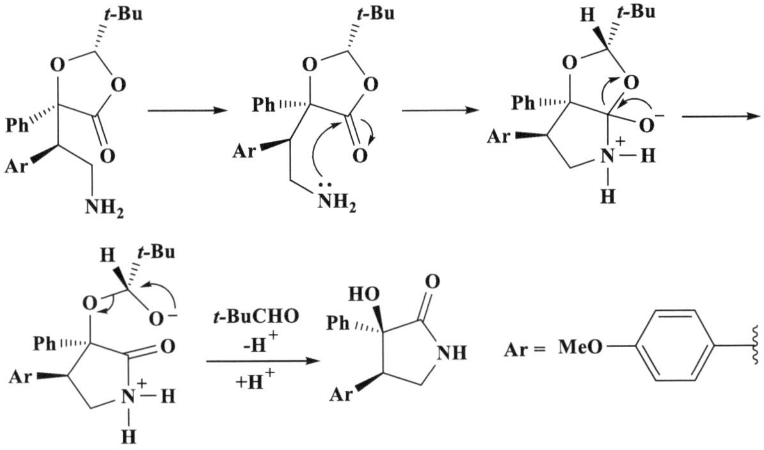

Scheme 1.2

No carbamate-protected analogues had been described in the literature [13]. The N-Boc-piperidone (commercially accessible on bulk scale) was transformed to its oxime and further oxidized with trifluoroacetic anhydride/hydrogen peroxide [14] to afford the 4-nitropiperidine in 55% yield. Under mild conditions, the nitro ester intermediate was prepared by a Michael addition of nitro piperidine to methyl acrylate. The amino intermediate, obtained by the reduction of NO_2 group by catalytic hydrogenation over Raney-Ni, was cyclized to spirolactam. The formation of nitro ester was more difficult, required tetra-n-butylammonium fluoride as a base and increased reaction time in refluxing tetrahydrofuran. Both spirolactams (Scheme 1.3) were formed in good overall yield (more than 80% over three steps) [15].

Innovative effort was described by Li et al. [16] in which the adduct was transformed to synthetically beneficial chiral Δ^1-pyrrolidine by an easy hydrogenation reaction (Scheme 1.4).

Developing an efficient and selective catalytic procedure was the best solution when asymmetry arose from prochiral starting compounds. The synthesis of endothelin-A antagonist ABT-546 was an excellent instance [17,18]. Although racemic pyrrolidine core of ABT-546 was constructed by developing a synthetic pathway (Scheme 1.5), it needed subsequent resolution with D-tartaric acid. This procedure still needed many recrystallizations and the resolved tartrate salt was formed in only 40% of theory. This inspired a search for an asymmetric synthesis of pyrrolidine core to afford the sufficient material to support upcoming scientific trials [19].

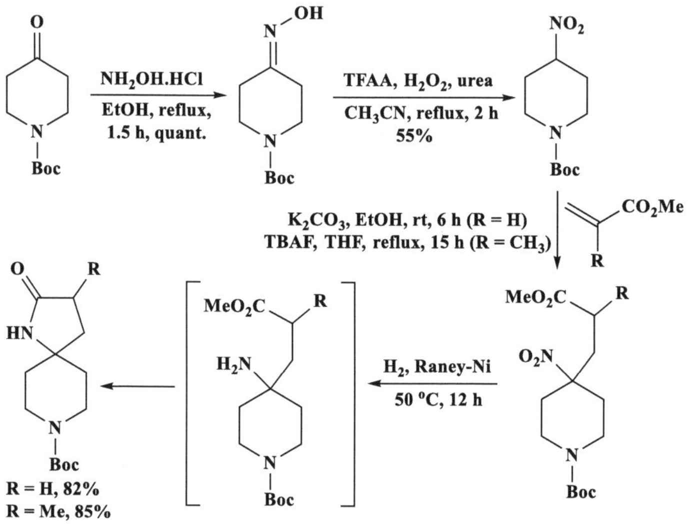

Scheme 1.3

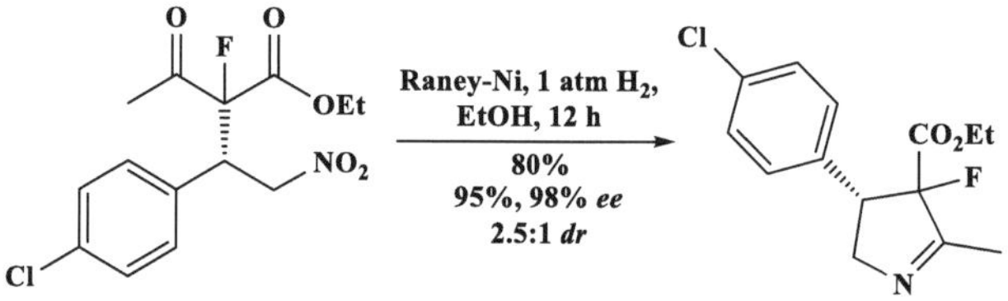

Scheme 1.4

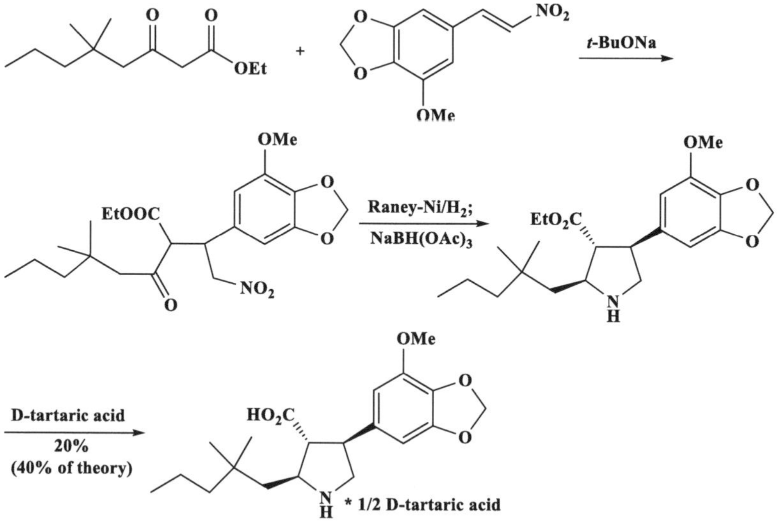

Scheme 1.5

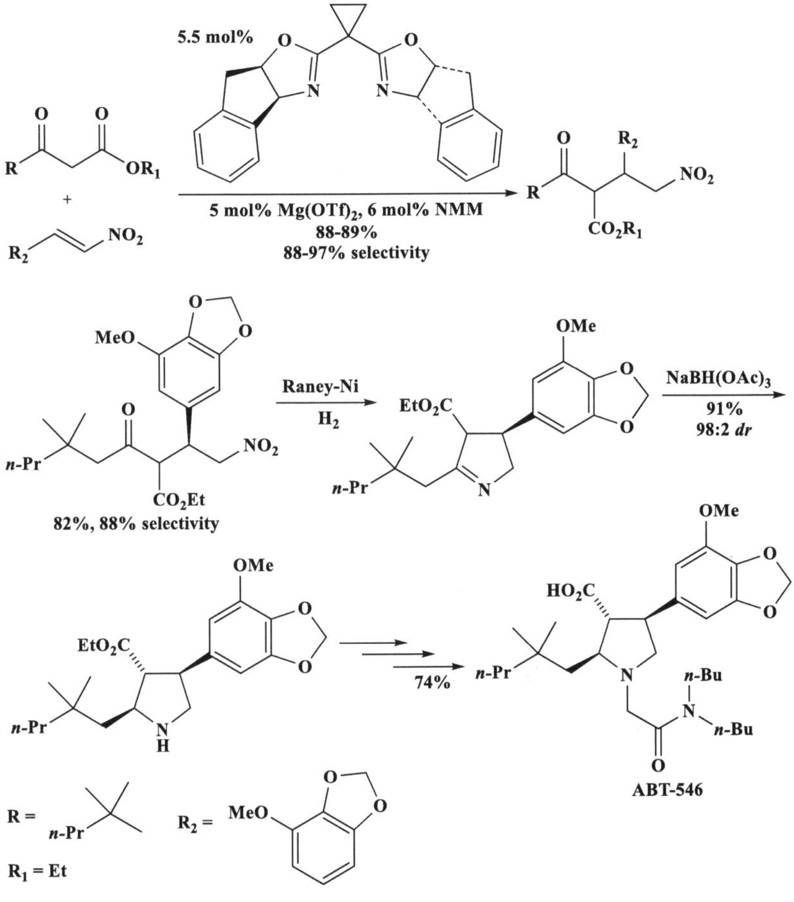

Scheme 1.6

Easily enolizable compounds like β-ketoesters were reacted with nitroalkenes using NMM (N-methylmorpholine) and catalytic amounts of chiral bis-oxazoline-magnesium triflate compound as cocatalyst (Scheme 1.6) [17]. The β-H residue present in adduct was too acidic to allow configurational stability at the corresponding carbon, on the other hand, the selectivity was good at the other stereocenter. This process was amenable to scaling up to 13 moles and was used for the formation of endothelin A antagonist ABT-546, (R)-rolipram, and phosphodiesterase-type 4 IC86518 [18,19]. The final compound was synthesized by reacting an appropriate β-ketoester with nitroalkene to give the desired adducts with 88% selectivity. The reduction of NO_2 group further provided pyrroline along with parent

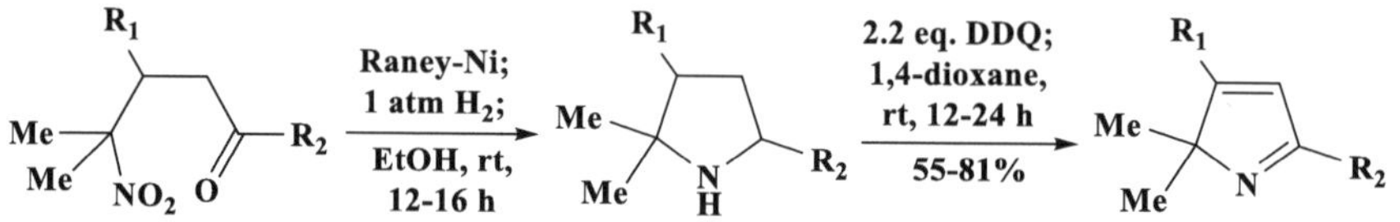

Scheme 1.7

nitrone in smaller amounts. The stereoselective reduction of this mixture was best effected with NaBH(OAc)$_3$ under acidic conditions, providing a pyrrolidine system which was transformed to ABT-546 by easy synthetic operations.

The reductive cyclization of substrates, prepared by conjugate addition of 2-nitropropane to α,β-unsaturated ketones, provided pyrrolidines which were transformed to 2H-pyrroles by dehydrogenation with DDQ (2,3-dichloro-5,6-dicyano-1,4-benzoquinone) (Scheme 1.7) [20,21].

The first deliberately used nitro-Mannich reaction was published, when Tsuritani et al. [22] described the formation of a strong κ-opioid agonist, i.e., ICI-199441 [23]. The synthesis started with nitro-Mannich reaction of N-phosphinoyl imine and CH$_3$NO$_2$ in the presence of heterobimetallic catalyst. The pyrrolidine was synthesized upon hydrogenation over Raney-Ni followed by an intramolecular. Then, acidic deprotection of phosphinoyl group, reductive methylation, and acylation delivered ICI-199441 in 35% overall yield and only six steps (Scheme 1.8).

The diastereoselectivity was not improved by adding 20 mol% (S)-proline [24] or the chiral amine organocatalysts developed by MacMillan et al. [25,26]. Therefore, this pathway was not further pursued, although disubstituted nitroalkanes afforded highly functionalized pyrrolidine building blocks after hydrogenation and deprotection (Scheme 1.9).

The use of enantioselective decarboxylative reaction of MAHTs (malonic half thioesters) for the formation of therapeutic targets was exemplified by the synthesis of GABA receptor antagonists utilizing organocatalysts (Scheme 1.10). The γ-nitrothioesters easily accessible via these organocatalytic methodologies occurring under mild conditions and tolerating both air and moisture are versatile building blocks for additional modifications. Among them the synthesis of γ-butyrolactams by the reduction of NO$_2$ group followed by an intramolecular cyclization led to intermediates for the synthesis of antidepressant (R)-rolipram [27] and to gram scale formation and transformation to (S)-baclofen·HCl, a GABA receptor antagonist utilized to cure spasticity [28,29].

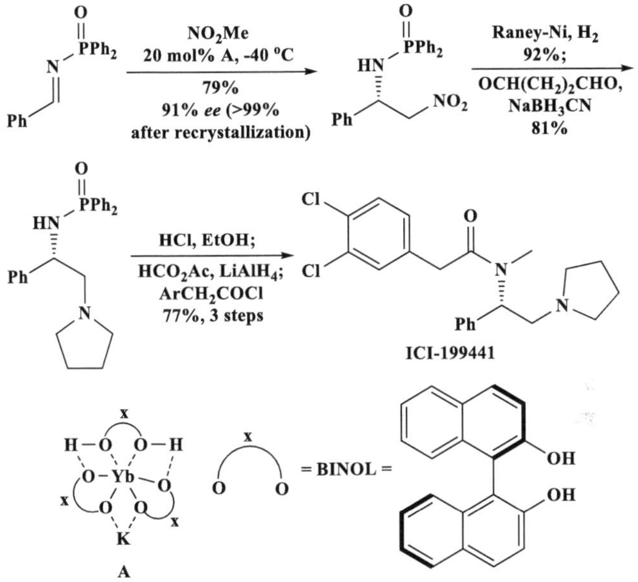

Scheme 1.8

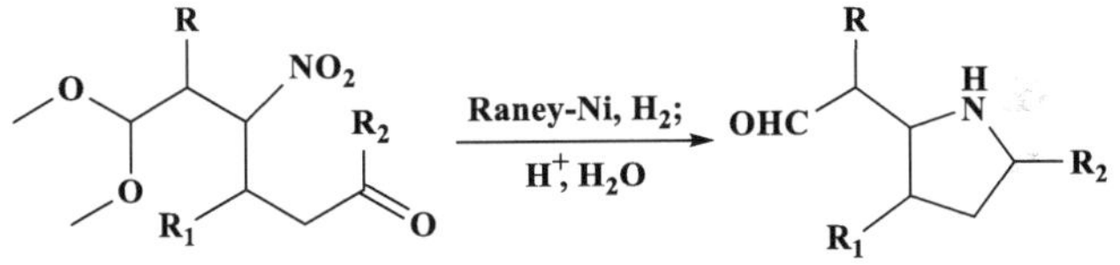

Scheme 1.9

An alternative metal-catalyzed system [18] where the potential for scale-up was clear is depicted in Scheme 1.11 and seemed to be highly competitive with an organocatalyzed method. The chiral Lewis acid-catalyzed Michael addition of diethyl malonate to nitrostyrene provided nitro ester, which on reduction and saponification led to desired rolipram. Both enantiomers of rolipram were obtained in total six steps and at 10 g scale with excellent overall yields (76%) and without chromatography [29].

The azabornane skeleton is present in a number of compounds, i.e., P antagonists and a common approach for its formation included an intramolecular ring-closure of appropriate derivatives of pyrrolidine. Tandem double conjugate additions of 4-benzylaminoenoate with nitroalkenes afforded 3-nitropyrrolidines, featuring a 3,4-*cis*-relationship (Scheme 1.12)

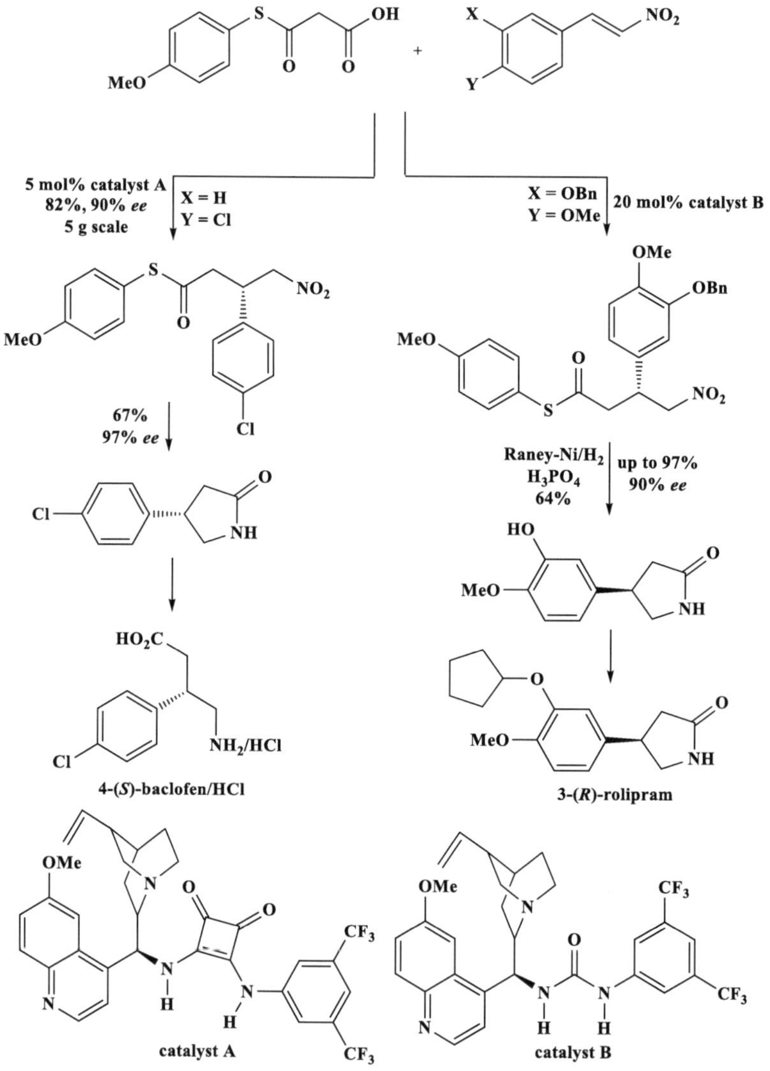

Scheme 1.10

[30]. The 3,4-*trans*-isomers were prepared smoothly by equilibration of 4-benzylaminoenoate under basic conditions. The reduction of ester and NO$_2$ groups via reaction with different reagents provided amino alcohols, which were subsequently protected with N-Boc. The concurrent nucleophilic displacements occurred on the mesylation of OH groups to produce the NH$_4^+$ salts of azabornane systems, which gave free amino derivatives by hydrogenolytic debenzylation.

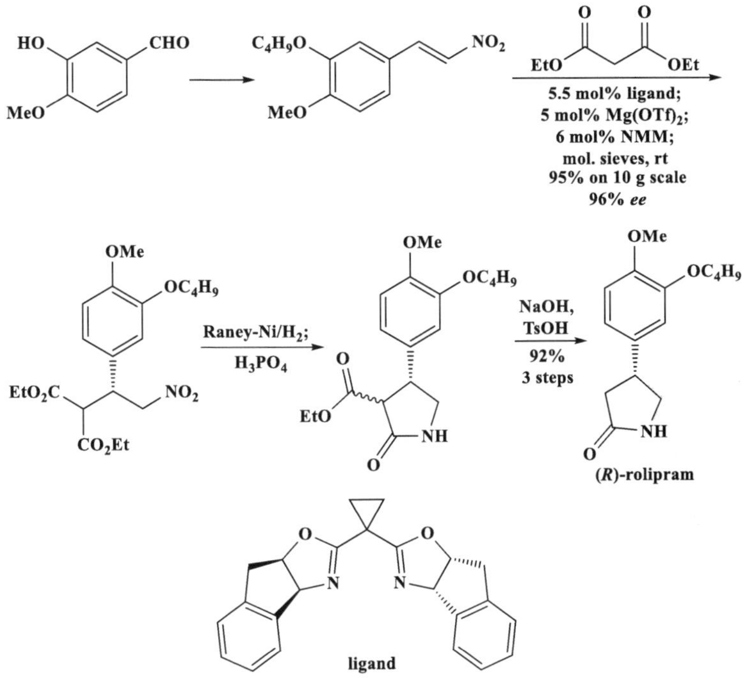

Scheme 1.11

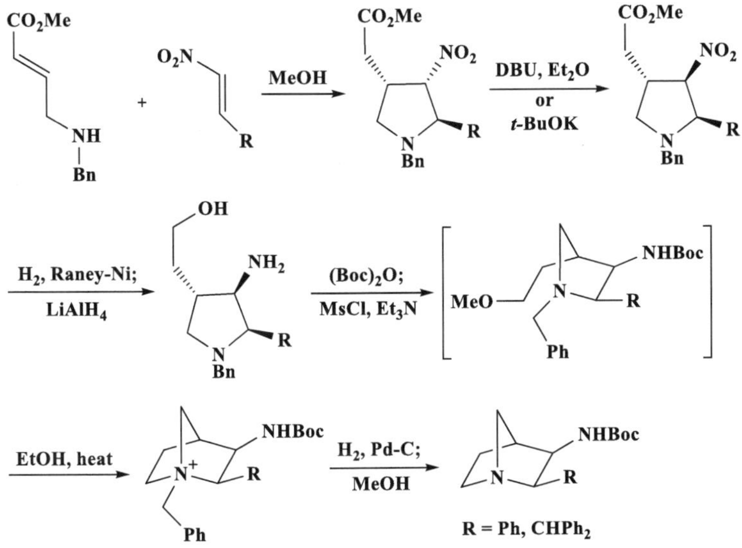

Scheme 1.12

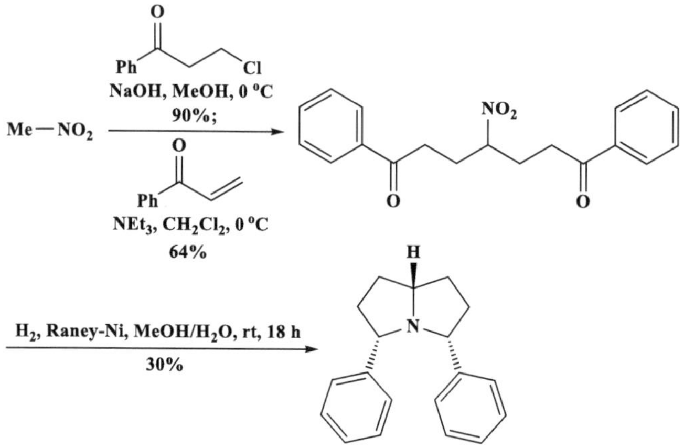

Scheme 1.13

1.3 Nitro group reduction involved in the synthesis of fused pyrroles

The CH$_3$NO$_2$ afforded *meso*-diphenylpyrrolizidine in three steps (Scheme 1.13). The CH$_3$NO$_2$ was reacted with sodium hydroxide, and the formed anion was utilized to displace the chloride from 3-chloro-1-phenylpropan-1-one to provide the nitroketone in 90% yield. Two-directional formation of diketone in this manner not proved to be feasible; however, it was obtained in good yield by Michael addition of nitroketone anion to phenylvinylketone. Exposing diketone to hydrogen gas and Raney-Ni further reduced the NO$_2$ group and affected the double reductive amination to provide the *meso*-diphenylpyrrolizidine in 30% yield. It is probable that the scope of this sequence can be extended to include quinolizidine and indolizidine frameworks, and therefore offered an alternative pathway to these scaffolds, instead of the double Michael addition approach [31].

The selection of Lewis acid promoter for these reactions changed the sense of asymmetric induction [32–36]. For instance, tandem [3+2]/[4+2]-cycloadditions (Scheme 1.14) assisted by Ti(O*i*-Pr)$_2$Cl$_2$, followed by hydrogenolysis provided tricyclic (-)-α-hydroxy lactam with 98% enantiomeric excess and use of MAPh (methylaluminium-bis-2,6-diphenylphenoxide) in the similar reaction provided (+)-α-hydroxy lactam with 93% enantiomeric excess. The chiral auxiliary did not affect the selectivity. Rather, it was due to a highly *endo*-selective cycloaddition in the case of titanium as compared to high *exo*-selectivity in the case of methylaluminium-bis-2,6-diphenylphenoxide [5].

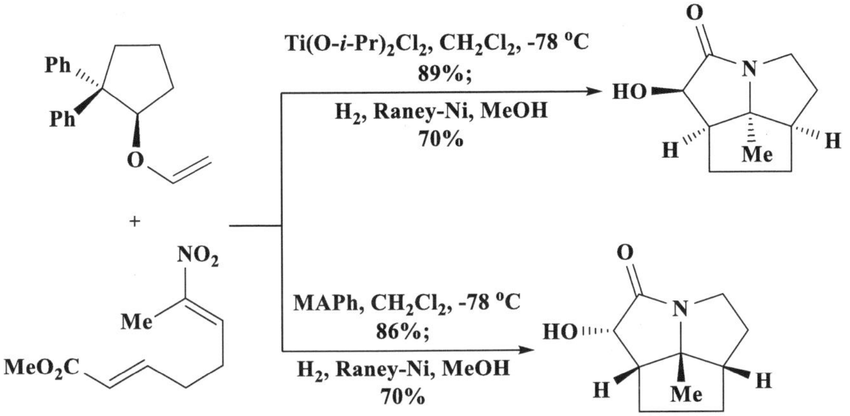

Scheme 1.14

The α-hydroxy lactams were obtained in nonracemic form by tandem cycloaddition with chiral vinyl and propenyl ethers (Scheme 1.15) [5, 37,38]. In this procedure, a standard set of reactions conditions was utilized, exposure of a solution of dienophile and nitroalkene to TiCl$_2$(Oi-Pr)$_2$ at -78°C, to give the nitronate intermediate. The thermal cycloaddition occurred over the period of 2–3 h at rt. The formed nitroso acetals were immediately subjected to hydrogenolysis with Raney-Ni in CH$_3$OH to give the lactams in 76% and 89% yield (three steps) with high enantioselectivity (96:4 enantiomeric ratio) [39]. The epimeric α-hydroxy lactam was easily obtained in 63% yield following the standard procedure (96:4 enantiomeric ratio).

The 7,8-dihydro-5(6H)-quinolone was used as a starting point for the formation of nicotine. An important characteristic was the use of nitroethylene, a synthon for constructing pyrrolidine ring of nicotine, as shown in Scheme 1.16. The reaction of 7,8-dihydro-5(6H)-quinolone with LDA (lithium di-i-propylamide) in tetrahydrofuran at -70°C followed by addition of nitromethylene afforded nitroketone in 63% yield. No dinitro side-product was formed. The reduction of nitroketone with H$_2$ (50 psi) and Raney-Ni in C$_2$H$_5$OH directly delivered bridged myosmine in 76% yield, apparently including an intermediate amino ketone which underwent an intramolecular Schiff base cyclization. The reduction of bridged myosmine with NaCNBH$_3$ in CH$_3$OH proceeded easily to give the bridged nornicotine in 53% yield. The proton nuclear magnetic resonance study of bridged nornicotine (at 80 and 500 MHz) showed the presence of only one of the two possible ring-juncture epimers. The reductive methylation

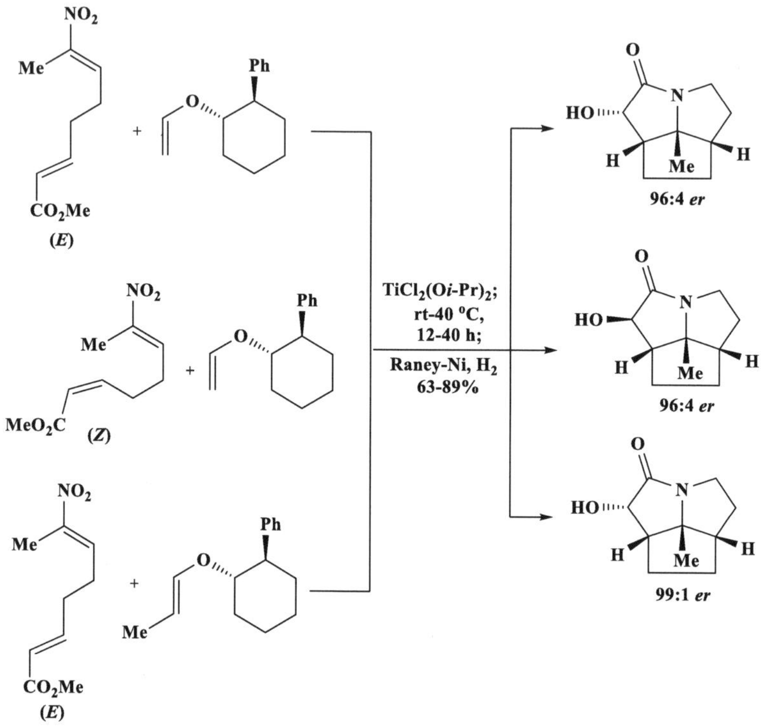

Scheme 1.15

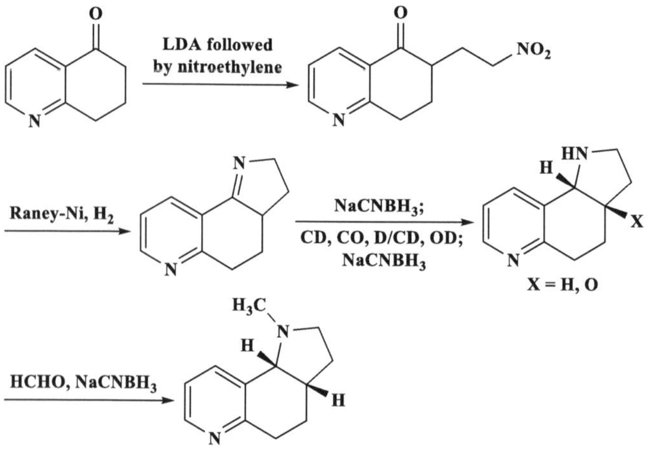

Scheme 1.16

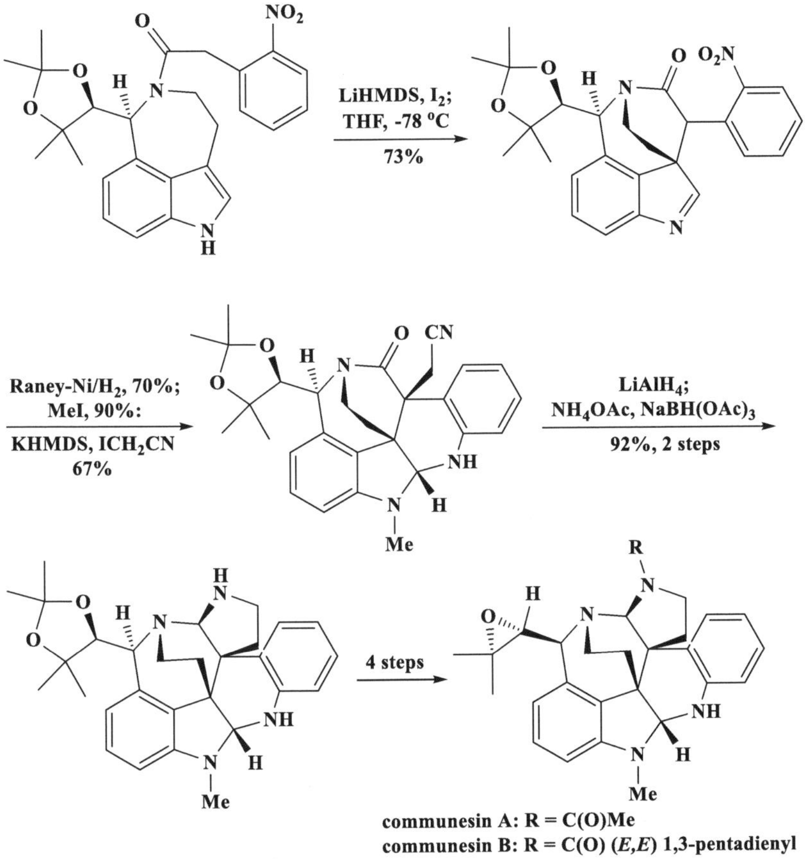

Scheme 1.17

of bridged nornicotine with NaCNBH$_3$ and aqueous HCHO in CH$_3$CN afforded bridged nicotine in 60% yield [40–45].

Zuo and Ma [46] described an elegant use of bridged lactams during the total synthesis of (-)-communesins A and B (Scheme 1.17). An oxidative cross-coupling of dianion formed from starting compound provided bridged lactam as a single diastereoisomer in 73% yield. The observed stereochemical outcome was proposed to arise from a favorable π-stacking interaction between indole and nitrophenyl rings. The lactam was advanced to cyanide, which underwent sequential reduction/reductive amination to give the aminal in 92% yield. This reaction most probably proceeded through the amino aldehyde intermediate; however, the mechanism involving a bridged iminium ion was not omitted. The aminal needed only four more steps to complete the formation of (-)-communesins A and B [6].

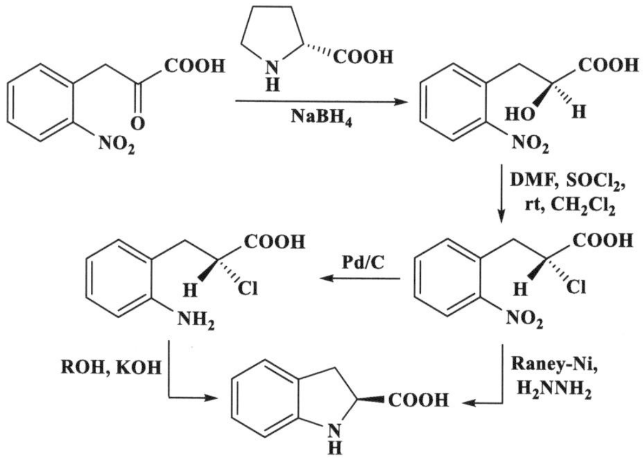

Scheme 1.18

1.4 Nitro group reduction involved in the synthesis of indoles

The enantiopure (S)-indoline-2-carboxylic acid was utilized as an intermediate in the formation of perindopril, i.e., a cardiotonic or antihypertensive active element. Additional beneficial uses of (S)-indoline-2-carboxylic acid were, for instance, in the formation of angiotensin-converting-enzyme inhibitors indolapril and pentopril. The enantiopure (S)-indoline-2-carboxylic acid and its methyl ester were obtained by traditional resolution, chemical formation via an asymmetric reduction with a chiral auxiliary or through the enzymatic approaches like resolution through the hydrolysis of indole-2-carboxylic esters or utilizing phenylammonia lyase for the formation of o-chlorophenylalanine followed by a Cu-catalyzed ring-closure. The drawback of traditional resolution process was the fact that the resolving agent was costly and difficult to recover. Additionally, the yield never exceeded 50%. The synthetic pathway via asymmetric reduction, utilizing a chiral auxiliary, included sodium borohydride reduction of prochiral 3-(o-nitrophenyl)pyruvic acid using chiral auxiliary D-proline. The resultant alcohol derivative was further transformed to enantioenriched (S)-indoline-2-carboxylic acid in 32% overall yield utilizing expensive D-(+)-proline (Scheme 1.18) [47].

The 2-(2-nitrophenyl)ethanol was prepared by hydroxymethylation of 2-nitrotoluene with p-formaldehyde under alkaline conditions (Scheme 1.19) [48,49]. In this analysis, Raney nickel was utilized as a catalyst both for the reduction of NO_2 group and for indole formation in the final

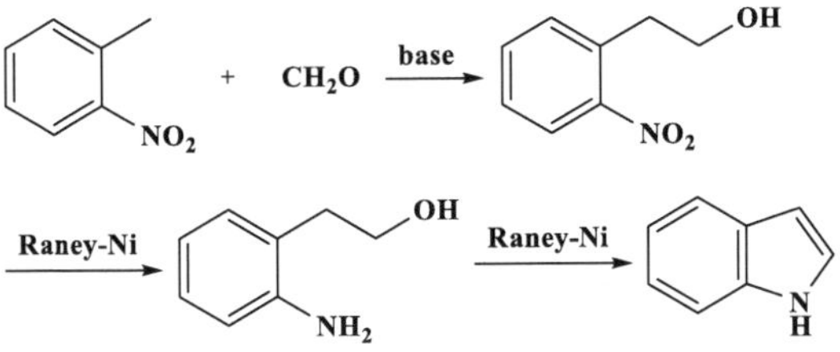

Scheme 1.19

step. The total yield for three consecutive steps was 78% depending on 2-NO$_2$C$_6$H$_4$CH$_3$ [7].

Having synthesized many *N*-substituted indole derivatives efficiently [50], then the desired indole was synthesized. The indole was synthesized from bromo compound following a seven-step procedure utilizing palladium/carbon-assisted indole formation as a key step (Scheme 1.20). The reaction of bromide with sodium sulfite delivered sulfonic acid, which on further treatment with phosphorus pentachloride delivered sulfonyl chloride derivative. Then, the reaction with aqueous methylamine sulfonyl chloride derivative afforded methanesulfonamide derivative, the NO$_2$ group of which was reduced with Raney-Ni to synthesize the aniline derivative. The iodination of aniline derivative utilizing elemental iodine delivered *o*-iodoaniline derivative, which was further transformed to methanesulfonamide, which upon reaction with phenylacetylene using 10% palladium/carbon–copper(I) iodide- triphenylphosphine and HOCH$_2$CH$_2$NH$_2$ in H$_2$O afforded indole derivative.

A sequence including Leimgruber–Batcho reaction was used to synthesize the 6-chloro-5-fluoroindole from 2-fluoro-4-methylaniline (Scheme 1.21) [51]. The nitration of C$_6$H$_5$NH$_2$ provided nitrotoluene, which was further reacted with NaNO$_2$ and CuCl to afford the chlorofluoronitrotoluene. The nitrotoluene was reacted with DMF-DMA to provide the nitroenamine, which underwent reductive cyclization with Raney-Ni and NH$_2$NH$_2$ to afford the 6-chloro-5-fluoroindole. The utilization of KI instead of CuCl provided 5-fluoro-4-iodo-2-nitrotoluene, and this resulted in the formation of 5-fluoro-6-iodoindole. The iodoenamine was reductively cyclized utilizing Fe in CH$_3$COOH [52].

Garden et al. [53] reported a modification of Sandmeyer's technique using EtOH as a cosolvent. This modification proved to be beneficial in cases where the aniline derivative was insoluble in the traditional reaction matrix. Use of modified Sandmeyer procedure afforded 4,6-dibromoisatin,

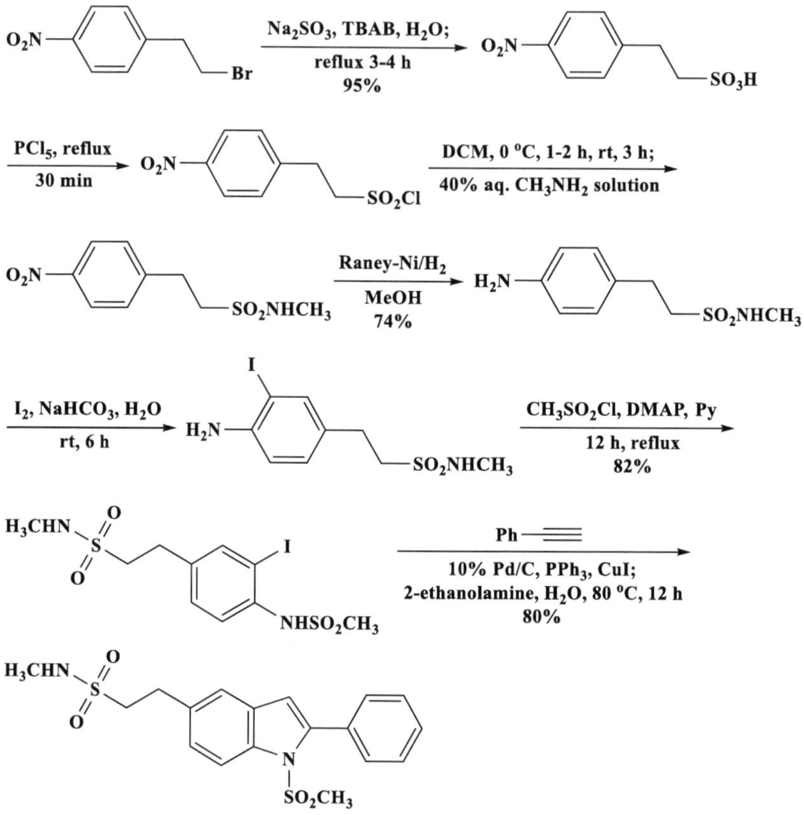

16   Raney nickel-assisted synthesis of heterocycles

Scheme 1.20

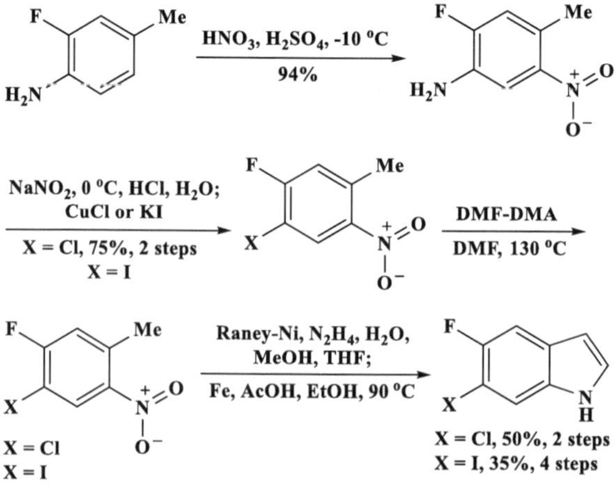

Scheme 1.21

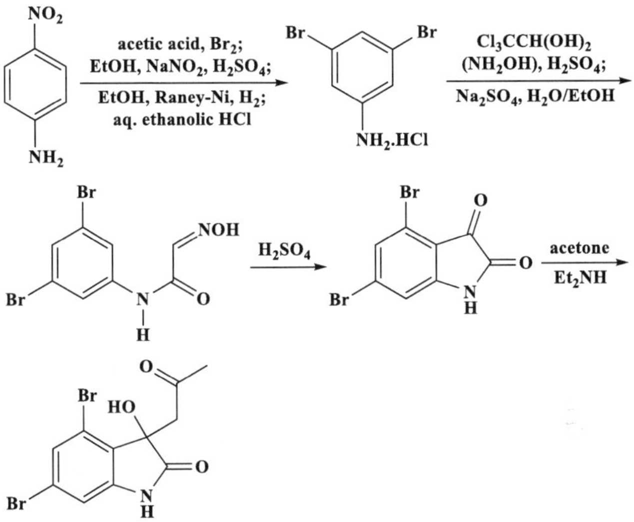

Scheme 1.22

which was an important intermediate for the formation of marine natural product 4,6-dibromo-3-hydroxy-3-(2-oxopropyl)indolin-2-one (convolutamydine A) in 88% yield (Scheme 1.22).

The *o*-nitrobenzylaryl sulfones, easily accessible through the vicarious nucleophilic substitution reaction of nitroarenes with carbanions of chloromethyl aryl sulfones, on reduction and transformation of the amino group into imidate [54,55] or imine [56] moiety, were able to able undergo cyclization to substituted indoles. This process was mainly beneficial because of the possibility to direct the vicarious nucleophilic substitution reaction selectively in *ortho*-position to the NO$_2$ group when the reaction was performed in potassium *t*-butoxide/tetrahydrofuran (Scheme 1.23) [57]. This method was utilized for the formation of 5- and 7-bromo-3-sulfonylindoles that were subsequently functionalized by Stille coupling with tributyl(vinyl)tin. The formed vinyl derivatives were then converted to amino compounds and were tested as norepinephrine reuptake inhibitors and 5-HT2A receptor antagonists [58,59].

The *o*-aminophenylacetaldehyde acetals and similar compounds were cyclized to indoles on heating with mineral acid. Generally, the 2,3-unsubstituted indoles were the products. The ring-closure occurred when semicarbazones of such aldehydes were hydrogenolyzed (Scheme 1.24) [60].

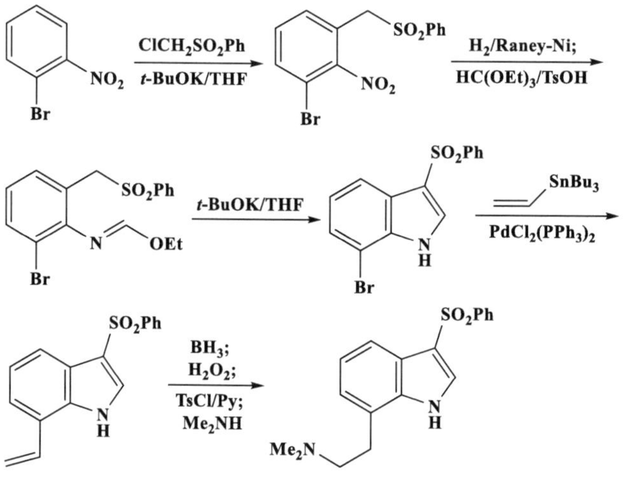

Scheme 1.23

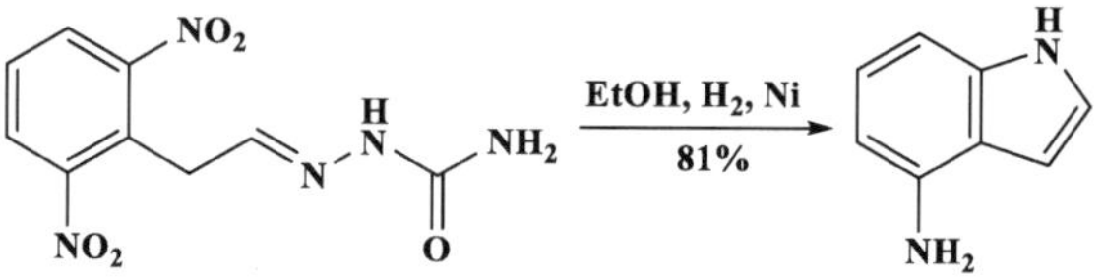

Scheme 1.24

1.5 Nitro group reduction involved in the synthesis of fused indoles

The successful efforts were made to prepare the 9-methoxyellipticine and ellipticine from accessible 6-bromo-5,8-dimethylisoquinoline (Scheme 1.25) [61]. The commercially accessible 4-methoxy-2-nitroaniline was heated with K_2CO_3, I_2, copper bronze, and isoquinoline to provide the coupled nitrobiarylamine in moderate yield. The reduction of NO_2 group with Raney-Ni and NH_2NH_2 in refluxing EtOH provided amine, which was not fully purified. The reaction of amine with $NaNO_2$ in HOAc provided triazole in an excellent yield (94%), calculated from nitrobiarylamine. The 9-methoxyellipticine was obtained in good yield when a solution of triazole was passed through a 500°C quartz tube. This method was useful for the synthesis of 9-methoxyellipticine and ellipticine themselves, but posed difficulties as a general process. The severe 500°C thermolysis limited the process to durable ellipticine derivatives, and highly derivatized 2-nitroanilines were difficult to prepare [62].

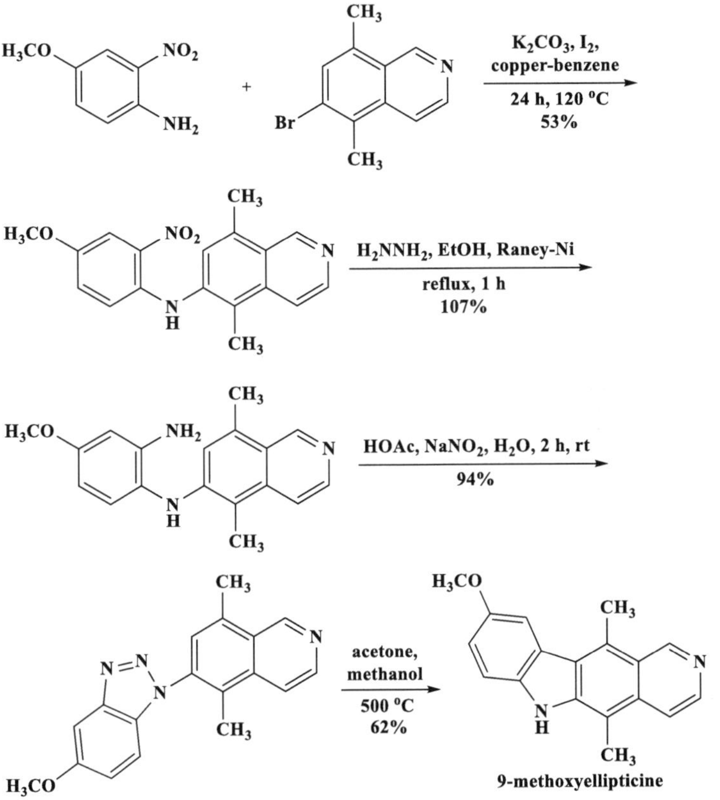

Scheme 1.25

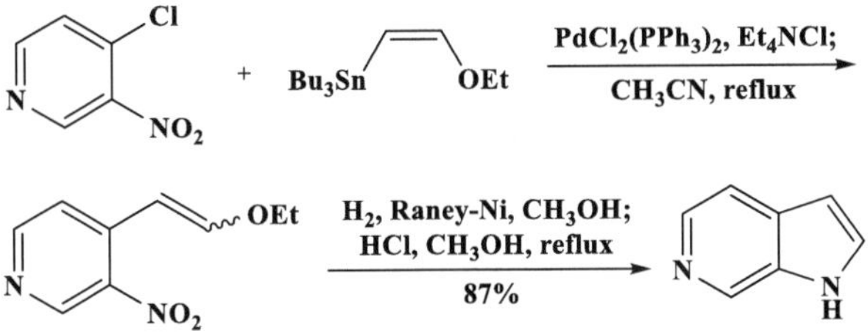

Scheme 1.26

Sakamoto et al. [63] described the Stille couplings of halonitro- and nitrotriflyloxypyridines with (Z)-1-ethoxy-2-tributylstannylethene utilizing 0.03 eq. bis(triphenylphosphine)palladium(II) chloride and 1 eq. tetraethylammonium chloride (TEAC) in CH$_3$CN to synthesize the nitrovinylpyridines (Scheme 1.26). In some examples both, the (E)- and

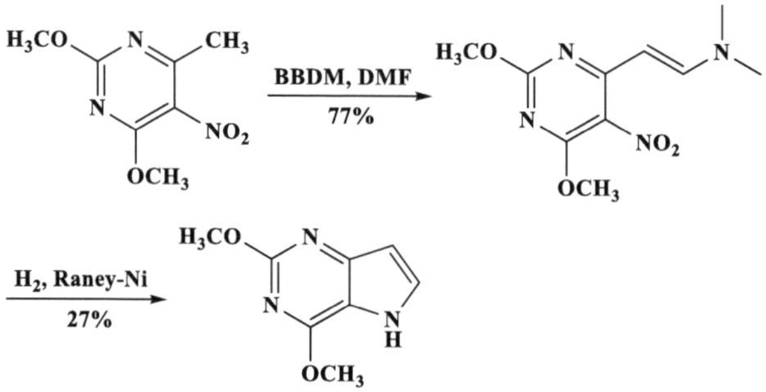

Scheme 1.27

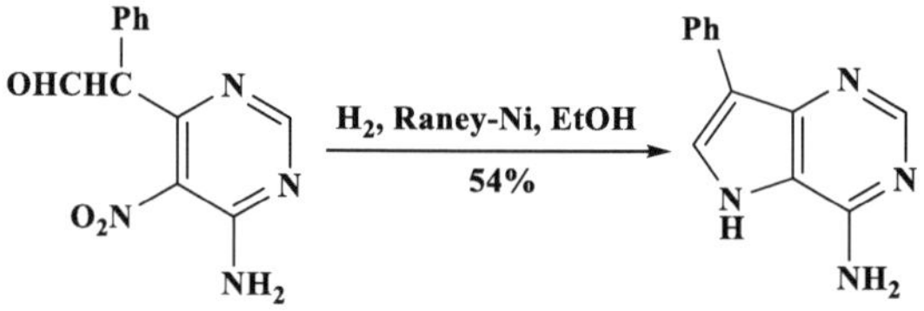

Scheme 1.28

(Z)-isomers were formed. Subsequently, the reduction of NO_2 functionality delivered aminopyridine, which was successfully cyclized in good yield under acidic conditions.

The starting compound was reacted with BBDM (*t*-butoxybis(dimethylamino)methane) to provide the cyclization precursor (Scheme 1.27). This was catalytically hydrogenated with Raney-Ni to afford the diazaindoles. The main disadvantage for both routes was low cyclization yield [64].

The catalytic hydrogenation of nitro(formylmethyl)pyrimidine was accompanied by cyclization and construction of a fused pyrrole ring (Scheme 1.28). The cyclization of 2-(2-aminophenyl)acetaldehyde acetals extended the technique to the formation of substituted indoles where the substituent was electron-withdrawing. This type of indole was not smoothly prepared by Bischler or Fischer synthesis [60].

1.6 Nitro group reduction involved in the synthesis of pyrazoles

The easiest modification in the amine was to replace the thiophenyl group by a bromo group (Scheme 1.29) because some of the desired amine, 7-

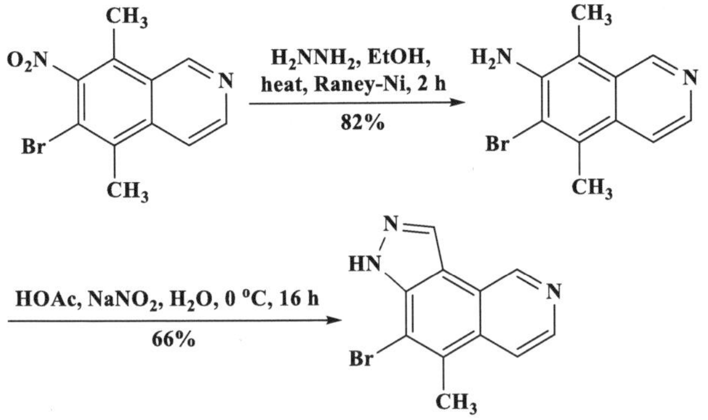

Scheme 1.29

amino-6-bromo-5,8-dimethylisoquinoline [65], was already in hand. The bromine has lone pairs like sulfur, but was deactivating rather than activating because of induction [66]. The reaction proceeded and displayed that the presence of an activating substituent was unnecessary at the 6-position. More amine was prepared from nitroisoquinoline by reduction with Raney-Ni and NH_2NH_2 in refluxing EtOH.

A common precursor was used for the generation of pyrazoloisoquinoline (Scheme 1.30). The 5-amino-4,7-dimethylindanone was oxidized to nitroindanone utilizing $NaBO_3$ in HOAc, but the yield was specified for crude material which was quite impure. Once purified, the indanone was reduced with $NaBH_4$ in cold MeOH to provide the indanol in excellent yield. An elimination with p-TsOH in refluxing C_6H_6 with the removal of H_2O provided 4,7-dimethyl-6-nitroindene in very good yield. Then, the reaction of indene with O_3, Me_2S, and aqueous NH_3 provided 6-nitro-5,8-dimethylisoquinoline in good yield. The reduction with NH_2NH_2 and Raney-Ni in refluxing EtOH provided a record yield of amine [67]. The diazotization of amine in cold HOAc provided pyrazoloisoquinoline in higher yield than the 6-aminoisoquinolies.

To determine if a substituent at the position-6 was required at all, the 7-amino-5,8-dimethylisoquinoline was synthesized (Scheme 1.31) beginning from a substantial gift of 4,7-dimethylindanone [68] already in hand. The reaction with cold fuming HNO_3 rapidly provided 4,7-dimethyl-6-nitroindanone, which was further reduced with excessive and adequate comfort by cold $NaBH_4$ in MeOH to provide the indanol. Then, the reaction with p-TsOH in refluxing C_6H_6, while eliminating H_2O, removed the alcohol to provide the indene. The reaction with O_3 followed by Me_2S

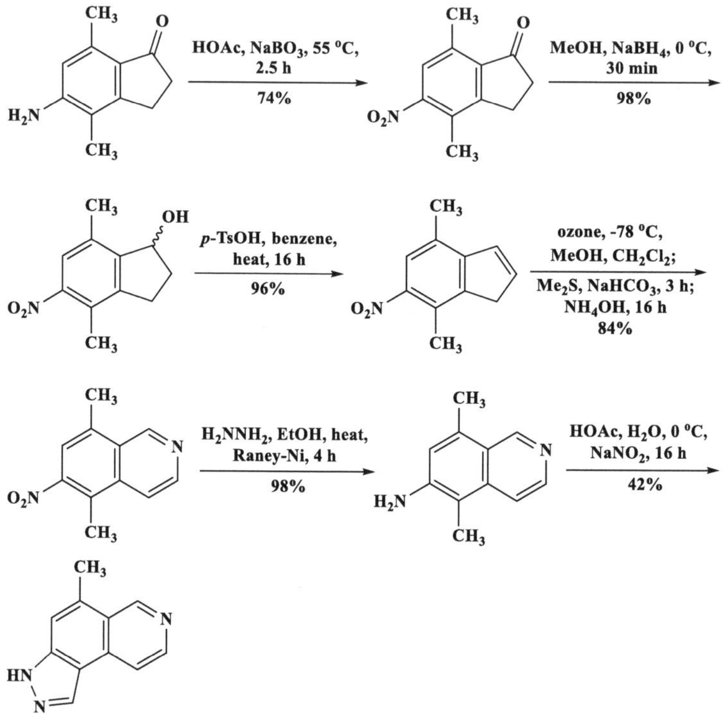

Scheme 1.30

and aqueous NH$_3$ provided 5,8-dimethyl-7-nitroisoquinoline easily. The nitroisoquinoline reduced unexpectedly rapidly with NH$_2$NH$_2$ and Raney-Ni in refluxing EtOH to provide the 7-amino-5,8-dimethylisoquinoline. The diazotization in cold HOAc indeed caused cyclization to 5-methyl-3H-pyrazolo[3,4-h]isoquinoline in good yield.

1.7 Nitro group reduction involved in the synthesis of imidazoles and benzimidazoles

Knudsen et al. [69] described that Raney-Ni-catalyzed hydrogenation of β-nitroamines afforded α,β-diaminoacid derivative. The reaction was carried out under much milder conditions to obtain the α,β-diaminoacid derivative in 80% yield without any loss of enantioselectivity (Scheme 1.32). Tsuritani et al. [22] also observed high yields utilizing this technique in their formation of ICI-199441.

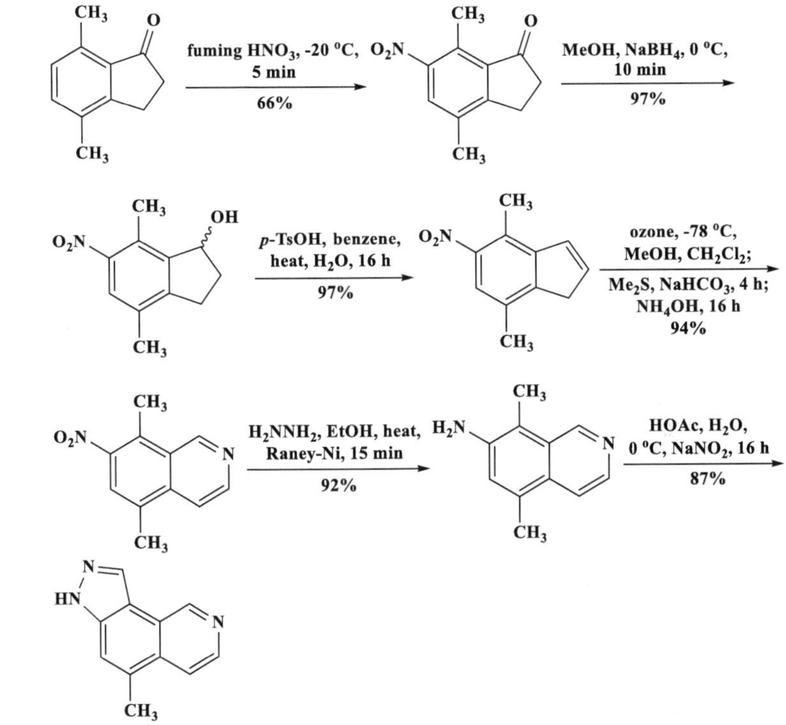

Raney nickel-assisted nitro group reduction for the synthesis of five-membered N-heterocycles 23

Scheme 1.31

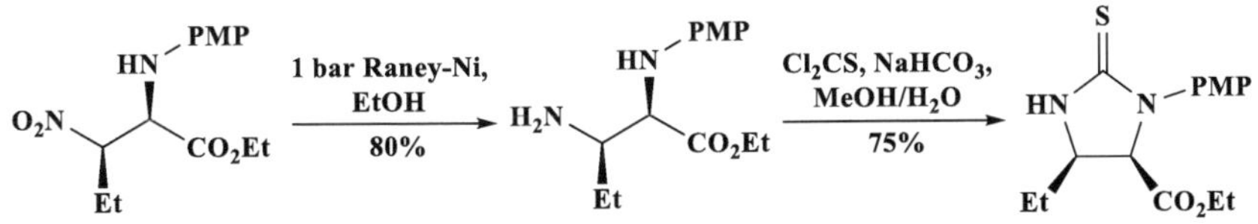

Scheme 1.32

The C$_6$H$_5$COOH was reacted with *m*-toluidine in NMP (*N*-methylpyrrolidinone) to give an intermediate. The NO$_2$ moiety was reduced with Raney-Ni and NH$_2$NH$_2$ in MeOH to provide the crude 3-amino-4-(*m*-tolylamino)benzoic acid, which was further transformed to benzimidazole by treating with catalytic *p*-TsOH and triethylorthoformate in tetrahydrofuran. This synthetic sequence was performed on a gram scale, and no flash chromatography or high performance liquid chromatography purification was essential to isolate any of the synthetic intermediates. The reaction of carboxylic acid with a variety of amines utilizing HATU and DIPEA (di-*i*-propylethylamine) in dimethylformamide delivered a library of 14 analogues of ML148 containing different amide functionalities. Disappointingly, none of these compounds bring about improved potency; however, it did deliver important information regarding the steric requirements around the cyclohexylamide moiety. A smaller ring or an exocyclic amine led to compounds with reduced activity (20–50 folds, respectively). The substituents at the position-3 on the piperidine ring were well tolerated, including the comparatively bulky *i*-propyl group. An addition of a 4-Br group only led to a moderate reduction in potency, the 4-trifluoromethyl group led to a severe loss of potency (80 fold) as compared to 3-trifluoromethyl group. The presence of heteroatoms (nitrogen or oxygen) at the position-4 was also not well tolerated, with activity in the micromolar range (*N*-methyl-piperazine) or inactivity (morpholine). The acyclic amides were very less active than cyclic ones, and secondary amides were less active than tertiary amides. As such, C_6H_5COOH was reacted with $(COCl)_2$ and a small amount of dimethylformamide in CH_2Cl_2, and the formed acid chloride was reacted with piperidine-HCl in CH_2Cl_2 to give the amide. The remainder of the reaction proceeded as planned, starting with the substitution of aromatic fluorine with many amines using di-*i*-propylethylamine or Hünig's base in dimethylformamide to deliver the anilines. The aromatic NO_2 moiety was reduced, and the obtained anilines were reacted with triethylorthoformate to give the target analogues. Analogues showed activities analogous to that of ML148 with comparatively flat structure–activity relationship being reported. An incorporation of small substituents in *para*- and in *meta*-position gave compounds with similar activities, while electron-withdrawing groups and bulkier substituents in *para*- and *meta*-position led to loss of potency. An introduction of nonaromatic groups had minimal influence on the potency (Scheme 1.33). These data proposed that this region could be subjugated for upcoming modulation of ADME properties since it seemed very tolerant to change. However, it remained unclear whether incorporation of a hydrophilic group will be tolerated since all analogues were lipophilic in this region [70].

A new series of 6-substituted benzimidazole-2-carbamic acid derivatives was prepared by treating substituted benzyl chlorides with 4-hydroxy-2-nitro aniline using anhydrous potassium carbonate in $(CH_3)_2CO$. A series of prepared 4-benzyloxy-2-nitro anilines was reduced with Ra-Ni as a catalyst. The 4-substituted 1,2-phenylenediamines were then treated with

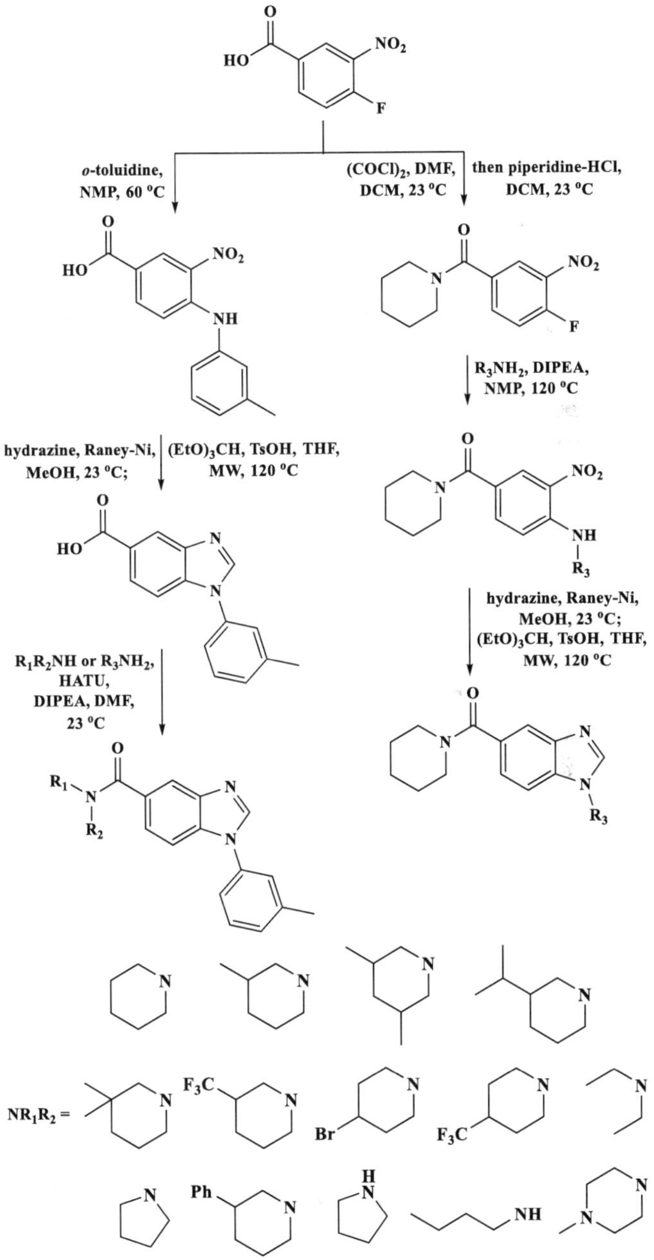

Raney nickel-assisted nitro group reduction for the synthesis of five-membered N-heterocycles 25

R₃ = phenyl, 2-OMe-phenyl, 3-Cl-phenyl, 3-OMe-phenyl, 3-CF₃-phenyl, 3-i-Pr-phenyl, 3-t-Bu-phenyl, 4-Me-phenyl, 4-Cl-phenyl, 4-OMe-phenyl, 4-CF₃-phenyl, 4-t-Bu-phenyl, t-Bu, t-Bu.

Scheme 1.33

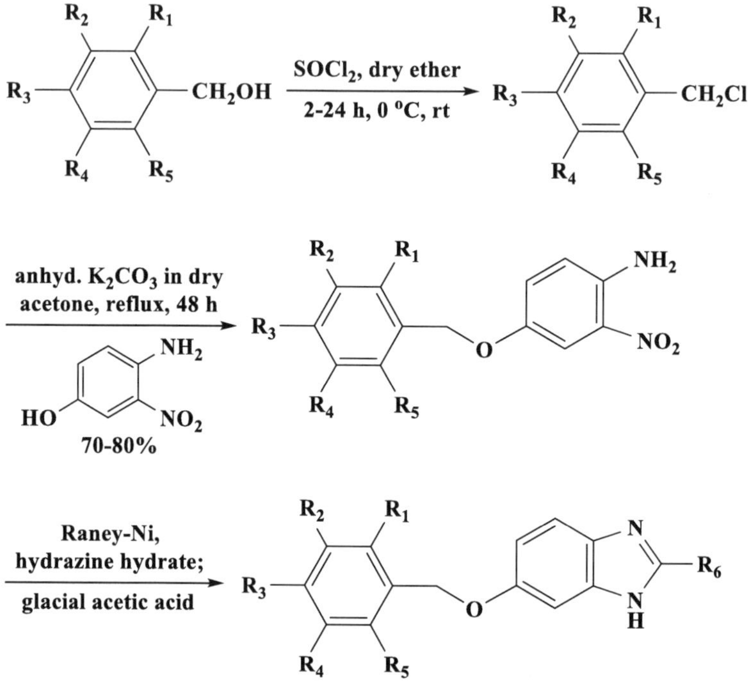

Scheme 1.34

$R_1 = R_5 = H, R_2 = R_3 = R_4 = OCH_3, R_6 = NHCOOCH_3$
$R_1 = R_5 = H, R_2 = R_3 = R_4 = OCH_3, R_6 = NHCOOCH_2CH_3$
$R_1 = R_2 = OCH_3, R_3 = R_4 = R_5 = H, R_6 = NHCOOCH_3$
$R_1 = R_2 = OCH_3, R_3 = R_4 = R_5 = H, R_6 = NHCOOCH_2CH_3$
$R_1 = R_4 = OCH_3, R_2 = R_3 = R_5 = H, R_6 = NHCOOCH_3$
$R_1 = R_4 = OCH_3, R_2 = R_3 = R_5 = H, R_6 = NHCOOCH_2CH_3$

1,3-bis(alkoxycarbonyl)-S-methyl isothiourea to afford the 6-substituted benzimidazole-2-carbamic acids (Scheme 1.34) [71].

The direct reductive reaction of carbanilic acid derivatives providing alkyl esters of 2-amino-1-benzimidazole carboxylic acids was of commercial value, since later type products were converted thermally into alkyl esters of benzimidazole-2-carbamic acids (Scheme 1.35) [72].

The formation and estimation of new head-to-head bis-benzimidazole compound 2,2-bis[4'-(3"-dimethylamino-1"-propyloxy)phenyl-5,5-bi-1 H-benzimidazole was attempted (Scheme 1.36) [73]. The X-ray crystallographic analysis of a compound with deoxyribonucleic acid dodecanucleotide sequence d(CGCGAATTCGCC) displays the compound bound in the A/T minor region of a B-deoxyribonucleic acid duplex and that the

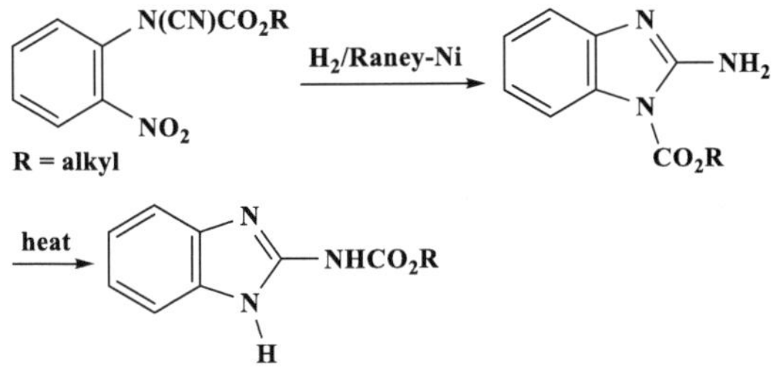

Scheme 1.35

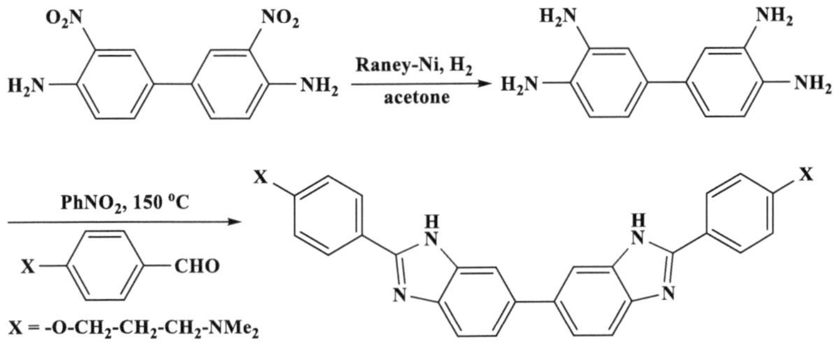

Scheme 1.36

head-to-head bis-benzimidazole motif hydrogen bonds to the edges of all four consecutive A:T base pairs. The compound exhibited potent growth inhibition with a mean IC$_{50}$ across an ovarian carcinoma cell line panel of 0.31 μM, with no significant cross-resistance in two aquired cisplatin-resistant cell line and a low level of cross-resistance in the P-glycoprotein over expressing acquired doxorubicin-resistant cell line. Studies with the hallow fiber assay and the in vivo tumor xenografts displayed.

An improved, convergent, and industrially beneficial procedure appropriate for the large-scale preparation of 2-mercapto-5-difluoromethoxy-1H-benzimidazole was needed for the formation of pantoprazole. The N-[4-(difluoromethoxy)phenyl]acetamide, prepared by the reaction of difluoromethylene chloride with 4-hydroxyacetanilide, underwent nitration followed by hydrolysis, reduction, and cyclization to afford the 2-mercapto-5-difluoromethoxy-1H-benzimidazole. In this procedure, the N-[4-(difluoromethoxy)phenyl]acetamide intermediate was synthesized by fluorination of 4-hydroxyacetanilide. Subsequent reactions such as nitration followed by hydrolysis, reduction, and cyclization

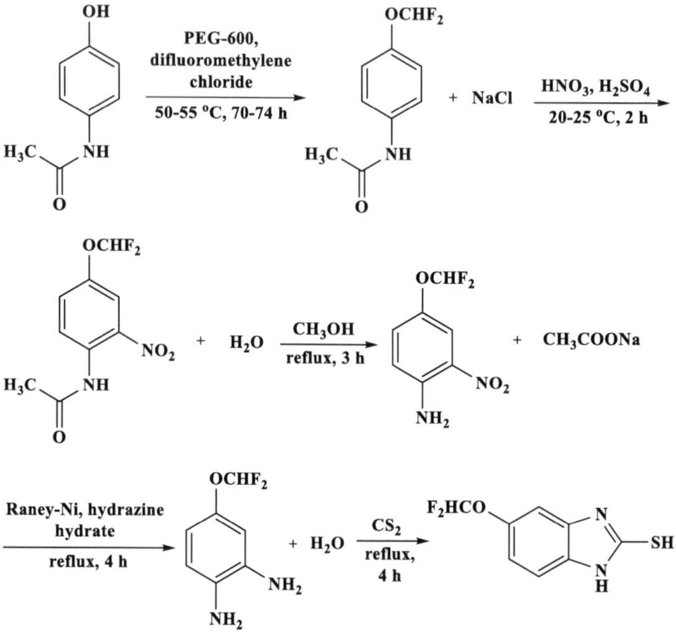

Scheme 1.37

of N-[4-(difluoromethoxy)phenyl]acetamide provided 2-mercapto-5-difluoromethoxy-1H-benzimidazole (Scheme 1.37). A better and convergent method for the formation of 2-mercapto-5-difluoromethoxy-1H-benzimidazole has been established using in situ procedure of N-[4-(difluoromethoxy)phenyl]acetamide giving 52.59% yield. The benefits were decreased cost, easy operation, good quality, and yield, the complete technical condition was mild and period was short and it was more appropriate for industrial synthesis [74].

The purines were synthesized from monoacyl pyrimidine-4,5-diamines. The 8-hydroxymethylpurinone was prepared from 5-acylamino-6-aminopyrimidin-4-one ethyl glycolate but not from keto ester and diamine. On the other hand, the mono(chloroacetyl)aniline underwent simultaneous cyclization and hydrogenolysis when NO_2 moiety was reduced catalytically. In this example, an o-acylaminoalkoxyamine was cyclized by heating with diethoxymethyl acetate to provide the purine (Scheme 1.38) [60].

The reaction was started by nucleophilic aromatic substitution of fluorine in substituted fluoronitrobenzenes utilizing potassium carbonate and $MeNH_2$ in a biphasic system (dichloromethane/water) to afford the

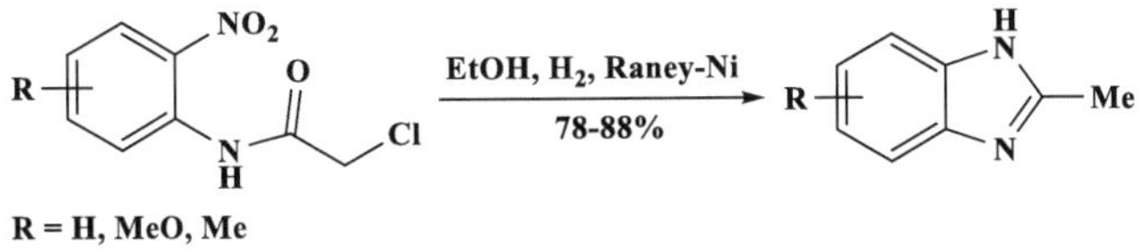

Raney nickel-assisted nitro group reduction for the synthesis of five-membered N-heterocycles | 29

Scheme 1.38

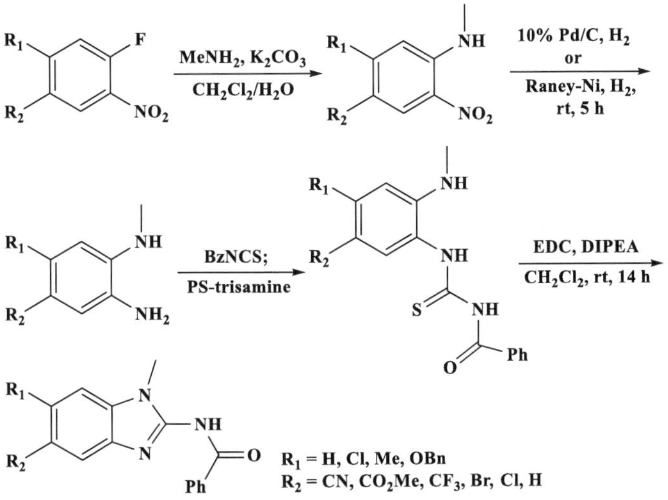

Scheme 1.39

nitroanilines (Scheme 1.39). It was remarkable to note that the nucleophilic aromatic substitution reaction of MeNH$_2$ with fluoronitro compounds proceeded very efficiently at room temperature and not needed forcing conditions or polar solvents like dimethylformamide, dimethylsulfoxide or N-methylpyrrolidone, which were usually utilized for solid-phase nucleophilic aromatic substitution displacement reactions. As a result, the nitroanilines were simply isolated after an aqueous work-up in quantitative yield with high purity (>95%), thus eliminating the need for any purification. The reduction of NO$_2$ moiety in nitroanilines was carried out with Ra-Ni/hydrogen gas (Scheme 1.40) or 10% palladium/carbon (Scheme 1.41) to synthesize the phenylenediamines, which were isolated in nearly quantitative yields by an easy filtration through a celite followed by solvent removal. While 10% palladium/carbon catalyst worked efficiently for most reductions, it resulted in complete hydrogenolysis of the Ar–Br bond in NO$_2$ compound to afford the dehalogenated phenylenediamine. This problem was not as serious as in the case of dichloro analog

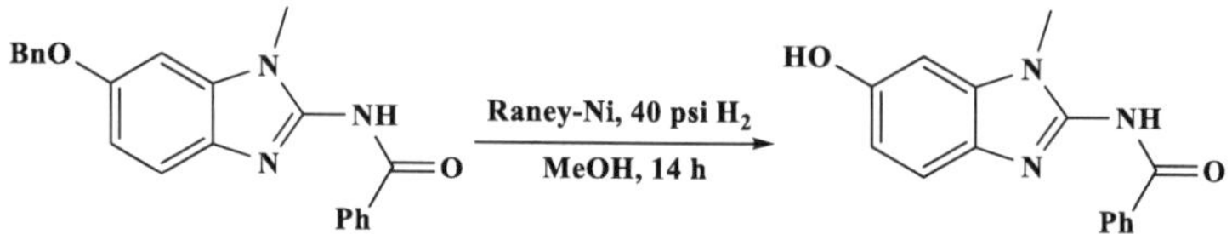

Scheme 1.40

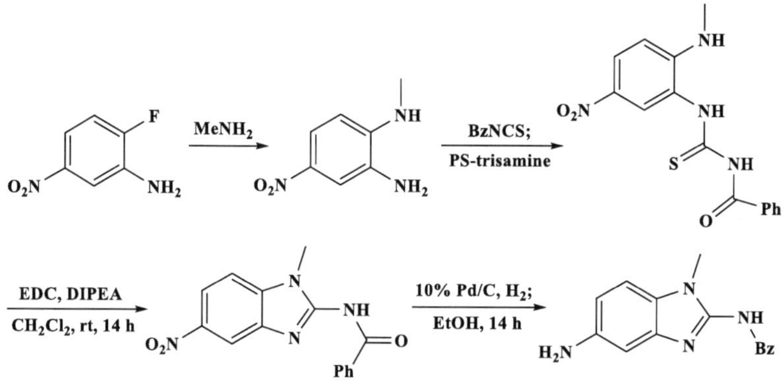

Scheme 1.41

where an employment of 10% palladium/carbon caused dehalogenation to a much smaller degree (<5%). The above mentioned problem was easily solved by applying Ra-Ni as an alternative of 10% palladium/carbon. Unexpectedly, the reduction with Ra-Ni was mild enough that the labile benzyl moiety in NO_2 compound survived the reaction. The reaction of diamines with a slight excess of benzoylisothiocyanate (1.1–1.2 eq.) in dichloromethane delivered benzoylthioureas. The excess benzyl isothiocyanate in the reaction was scavenged by an introduction of a small amount of PS-trisamine resin and the reaction was purified by filtering off the resin. The benzoylthioureas were further cyclized utilizing EDC (1-[3-(dimethylamino)propyl]-3-ethylcarbodimide hydrochloride) to deliver the 2-(N-benzoyl)-aminobenzimidazoles. The presence of EWG (electron-withdrawing group) on the thiourea was critical for an efficient cyclization of benzoylthioureas to 2-(N-benzoyl)-aminobenzimidazole. The cyclization utilizing EDC was slow and low yielding in the absence of an electron-withdrawing group. The final products 2-(N-benzoyl)-aminobenzimidazoles were obtained in high purity and yield (90–95%) following an aqueous work-up and not needed any further purification. In some cases, where the final purities were not >95%, the products were

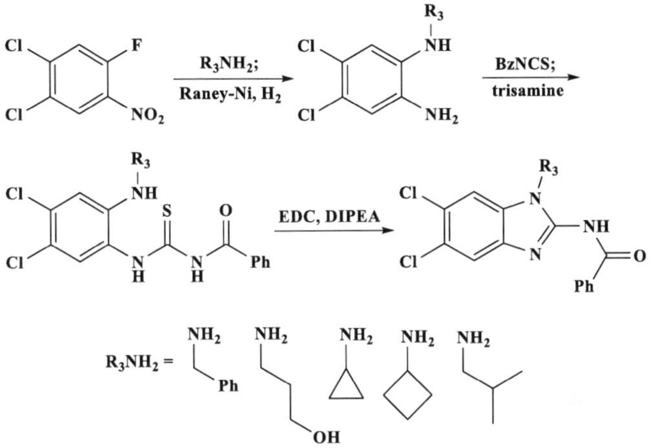

Scheme 1.42

smoothly purified by trituration with dichloromethane, hexanes or ether. In the case of aminobenzimidazole, the benzyl protecting moiety was cleanly removed utilizing Ra–Ni and H$_2$ gas (50 psi) to deliver the benzimidazole. The reduction of NO$_2$ group with 10% palladium/carbon afforded amino benzimidazole in quantitative yield [8].

The nucleophilic aromatic substitution reaction of starting compound with amines proceeded easily to give the nitroanilines, which were reduced with Ra–Ni and hydrogen gas at atmospheric pressure to afford the o-phenylenediamines. Then, treatment with benzyl isothiocyanate followed by removal of excess benzyl isothiocyanate utilizing trisamine resin afforded benzoylthioureas. The cyclization to 2-aminobenzimidazoles was performed utilizing 1-[3-(dimethylamino)propyl]-3-ethylcarbodiimide hydrochloride (Scheme 1.42) [8, 75–78].

In order to introduce the diverse alkylamino groups into benzimidazole 6-position, an intermediate having chlorine was generated by treatment with 2-(valeroylamino)-4-chloronitrobenzene (Scheme 1.43). The nucleophilic substitution of Cl atom with primary or secondary amines at 130°C provided amino compounds. The amino compounds were converted into benzimidazoles by reduction of NO$_2$ group followed by ring-closure in glacial CH$_3$COOH and cleavage of the ester moiety [9].

The benzimidazole was prepared by selective protection of 4-amino group in starting compound with phthalic anhydride followed by acylation, reduction of NO$_2$ group, and ring-closure in glacial CH$_3$COOH (Scheme 1.44) [9].

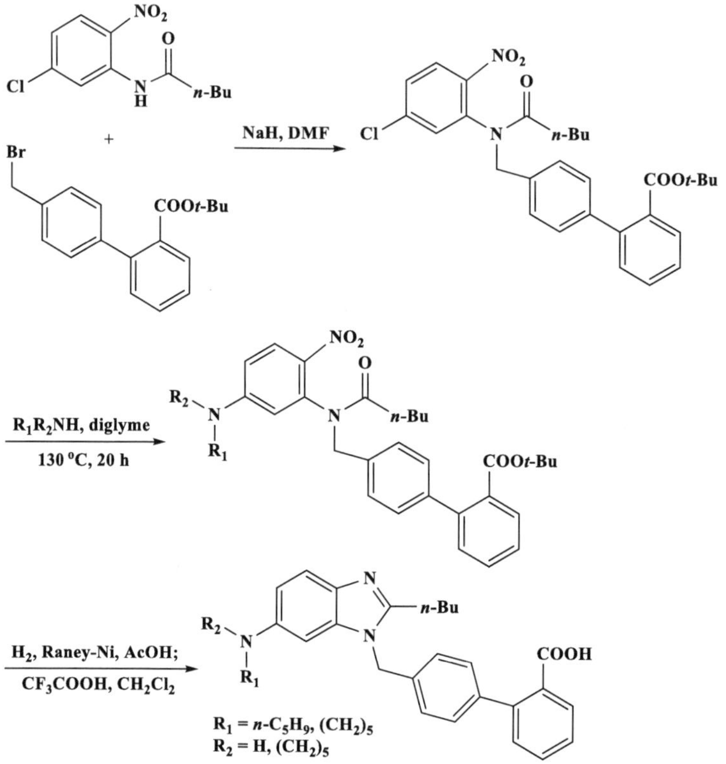

Scheme 1.43

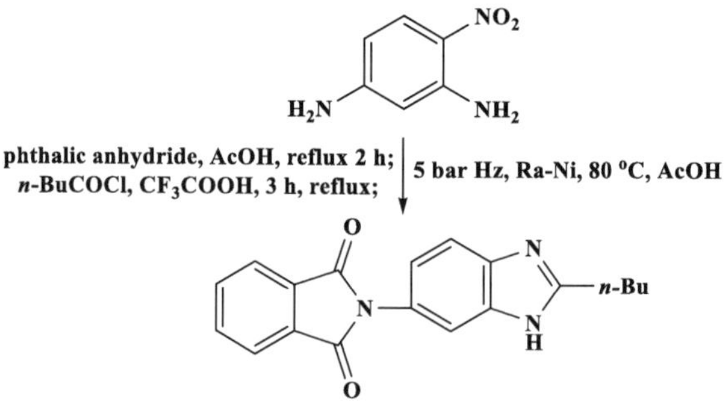

Scheme 1.44

The imidazopyridine-substituted benzimidazole was synthesized starting from substituted acetophenone (Scheme 1.45). The substituted acetophenone was prepared by an acylation of diethyl malonate with 4-(butyrylamino)-3-methylbenzoyl chloride followed by decarboxylation and

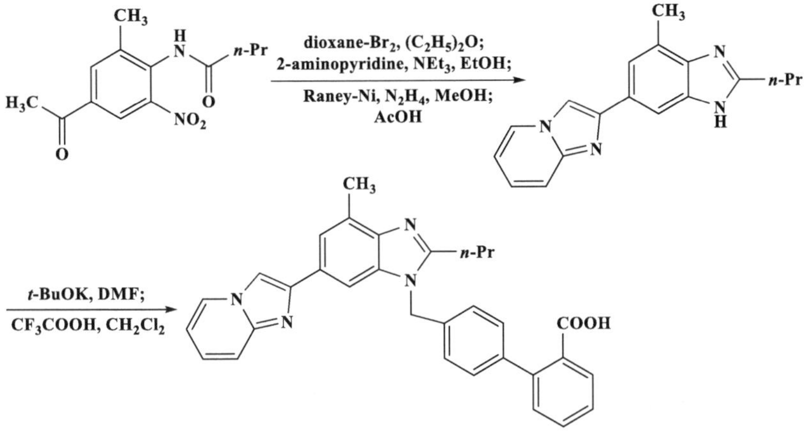

Scheme 1.45

nitration (not shown). The imidazopyridine-substituted benzimidazole was synthesized by side-chain bromination followed by condensation with 2-aminopyridine [79] and benzimidazole formation. The carboxylic acid was obtained after alkylation and ester hydrolysis [9].

Sharma and Pujari [80] synthesized thiazolo[3,2-a]benzimidazole-3(2H)-ones from 2,4-dibromo-6-nitroanilines (Scheme 1.46).

Ronne et al. [81] prepared various compounds with the aim of establishing a thorough structure–activity relationship analysis for such health hazardous products. The small-scale preparation was achieved by two routes starting from the same precursor. The first method was to react the in situ generated diaminoquinoline derivative directly with BrCN. Then, treatment with t-BuOOH provided methylated derivatives in 27–47% yield (method A). Another methodology was introduced to avoid the use of highly toxic, hazardous reagent BrCN according to the technique of Ziv et al. [82] but with some modifications. This multistep synthetic route was to synthesize the 2-mercaptoimidazo[4,5-f]quinoline as a center for amination by different pathways (methods B, C, and D) (Scheme 1.47) [10].

1.8 Nitro group reduction involved in the synthesis of triazoles

The C-derivatization of benzotriazole was somewhat problematic, and the benzotriazole system selectively substituted at the benzenoid ring was generally generated from scratch. An example of this case is shown in

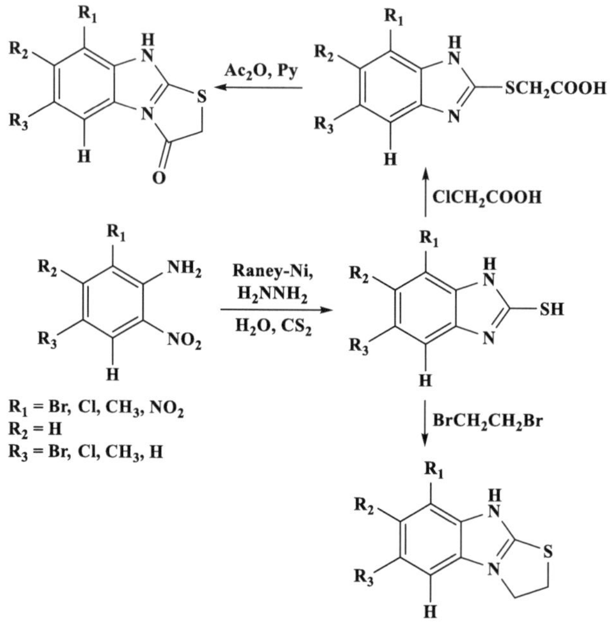

Scheme 1.46

Scheme 1.48. The procedure was started from a relatively simple molecule of 4-(2-chloroethyl)nitrobenzene, which was reduced to aniline and acetylated to provide the acetanilide. The chlorination with thionyl chloride occurred in 75% yield, which was subsequently nitrated to provide the product in 70% yield. The nitroaniline was prepared by the deprotection of amino group. In the following steps, the 2-chloroethyl substituent was transformed into 2-(dipropylamino) ethyl group, the NO$_2$ group was reduced, and the formed o-phenylenediamine underwent cyclocondensation with HNO$_2$ to provide the benzotriazole with the final step yield of 61% [83].

An ether linker was utilized for the reactions on the solid support, which was expected to be inactive towards various potential applications (Mannich reactions, displacement reactions including organometallic reagents and other nucleophiles, etc.). Therefore, a model sequence was examined in solution including 4-benzyloxy-1,2-phenylenediamine, 4-benzyloxy-1H-benzotriazole, and 4-benzyloxy-2-nitroaniline (Scheme 1.49). The 4-(benzyloxy)-2-nitroaniline was obtained when sodium salt of 4-amino-3-nitrophenol was treated with PhCH$_2$Cl in DMA (N,N-dimethylacetamide). The 4-(benzyloxy)-2-nitroaniline was subsequently reduced to 4-(benzyloxy)-1,2-phenylenediamine either by utilizing (i)

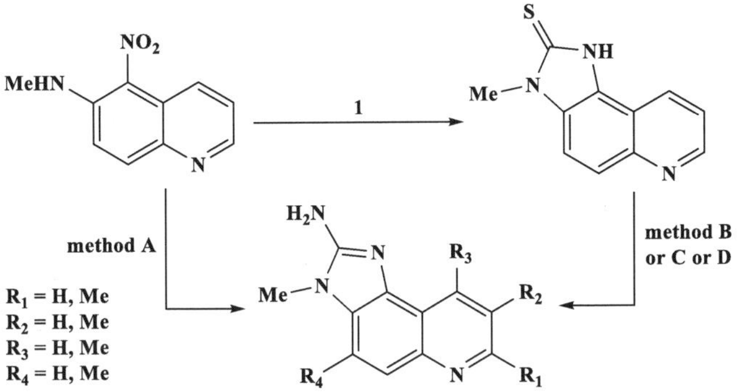

Reagents and conditions: (1) Na$_2$S$_2$O$_4$/MeOH/25% aq. NH$_3$/reflux/CS$_2$/MeOH/reflux, method A: (1) H$_2$, Raney-Ni/EtOH/rt/BrCN, (2) 70% aq. *t*-butylhydroperoxide/FeSO$_4$/1 M H$_2$SO$_4$/rt, method B: (1) MeI/reflux, (2) KMnO$_4$-AcOH/rt, (3) NaNH$_2$-liq. NH$_3$/reflux or NH$_3$-EtOH/150 °C, method C: (1) SOCl$_2$-POCl$_3$/reflux, (2) NaNH$_2$-liq. NH$_3$/reflux or NH$_3$-EtOH/150 °C, method D: (1) H$_2$O$_2$/rt, (2) NaNH$_2$-liq. NH$_3$/reflux or NH$_3$-EtOH/150 °C.

Scheme 1.47

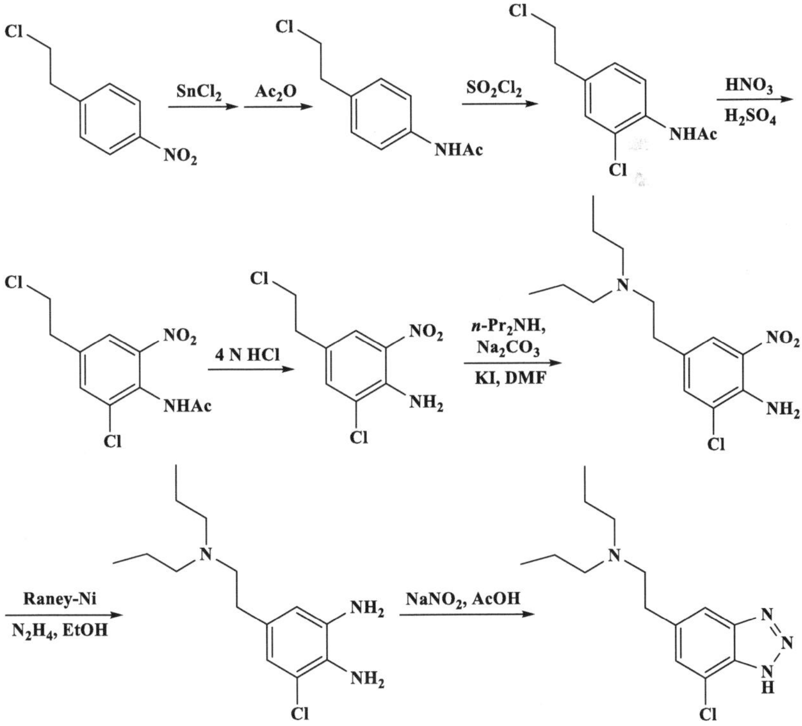

Scheme 1.48

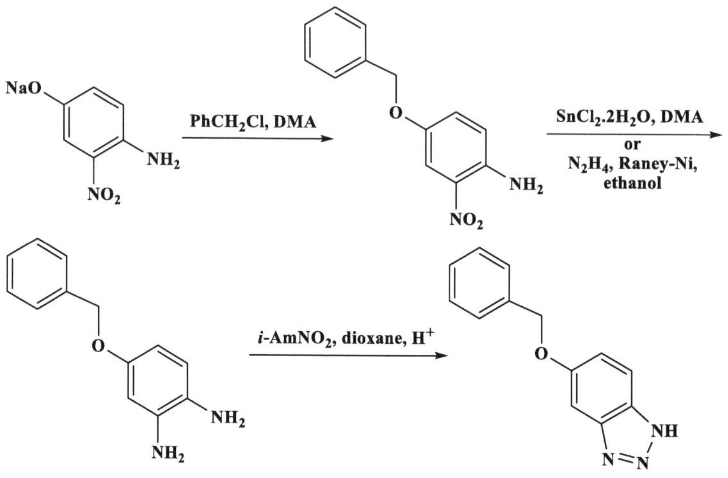

Scheme 1.49

SnCl$_2$ in *N,N*-dimethylacetamide or (ii) Ra-Ni and NH$_2$NH$_2$ in EtOH. The nuclear magnetic resonance spectra of both products were same, but it was difficult to remove the *N,N*-dimethylacetamide totally utilizing the first technique. The 4-(benzyloxy)-1,2-phenylenediamine products were both utilized without purification for subsequent reactions. The 5-(benzyloxy)-1*H*-benzotriazole hetero ring was constructed utilizing isoamyl nitrite in dioxane (in place of H$_2$O or mixed aqueous solutions); thus, this modification was used in the formation of a resin-bound benzotriazole nucleus [84].

References

[1] (a) Kaur N. Applications of microwaves in the synthesis of polycyclic six-membered *N,N*-heterocycles. Synth Commun 2015;45:1599–631; (b) Kaur N. Greener and expeditious synthesis of fused six-membered *N,N*-heterocycles using microwave irradiation. Synth Commun 2015;45:1493–519; (c) Kaur N. Review on the synthesis of six-membered *N,N*-heterocycles by microwave irradiation. Synth Commun 2015;45:1145–82; (d) Kaur N, Kishore D. Microwave-assisted synthesis of seven- and higher-membered *N*-heterocycles. Synth Commun 2014;44:2577–614; (e) Kaur N. Microwave-assisted synthesis of five-membered *S*-heterocycles. J Iran Chem Soc 2014;11:523–64; (f) Kaur N, Kishore D. Application of chalcones in heterocycles synthesis: synthesis of 2-(isoxazolo, pyrazolo and pyrimido) substituted analogues of 1,4-benzodiazepin-5-carboxamides linked through an oxyphenyl bridge. J Chem Sci 2013;125:555–60; (g) Kaur N, Kishore D. Metal and non-metal based catalysts for oxidation of organic compounds. Catal Surv Asia 2013;17:20–42.

[2] (a) Kaur N. Palladium-catalyzed approach to the synthesis of *S*-heterocycles. Catal Rev 2015;57:478–564; (b) Kaur N. Gold and silver assisted synthesis of five-membered

oxygen and nitrogen containing heterocycles. Synth Commun 2019;49:1459–85; (c) Kaur N. Synthesis of six- and seven-membered and larger heterocycles using Au and Ag catalysts. Inorg Nano Met Chem 2018;48:541–68; (d) Kaur N, Verma Y, Grewal P, Bhardwaj P, Devi M. Application of titanium catalysts for the syntheses of heterocycles. Synth Commun 2019;49:1847–94; (e) Kaur N, Bhardwaj P, Devi M, Verma Y, Grewal P. Photochemical reactions in five and six-membered polyheterocycles synthesis. Synth Commun 2019;49:2281–318; (f) Kaur N. Application of silver-promoted reactions in the synthesis of five-membered O-heterocycles. Synth Commun 2019;49:743–89; (g) Kaur N, Kishore D. Peroxy acids: role in organic synthesis. Synth Commun 2014;44:721–47.

[3] Yang SM, Malaviya R, Wilson LJ, Argentieri R, Chen X, Yang C, Wang B, Cavender D, Murray WV. Simplified staurosporine analogs as potent JAK3 inhibitors. Bioorg Med Chem Lett 2007;17:326–31.

[4] Schmidt AW, Reddy KR, Knölker HJ. Occurrence, biogenesis, and synthesis of biologically active carbazole alkaloids. Chem Rev 2012;112:3193–328.

[5] Denmark SE, Schnute ME, Senanayake CBW. Tandem inter [4+2]/intra [3+2] nitroalkene cycloadditions. 5. Origin of the Lewis acid dependent reversal of stereoselectivity. J Org Chem 1993;58:1859–74.

[6] Szostak M, Aube J. Chemistry of bridged lactams and related heterocycles. Chem Rev 2013;113:5701–65.

[7] Guo X, Peng Z, Jiang S, Shen J. Convenient and scalable process for the preparation of indole via Raney nickel-catalyzed hydrogenation and ring closure. Synth Commun 2011;41:2044–52.

[8] Seth PP, Robinson DE, Jefferson EA, Swayze EE. Efficient solution phase synthesis of 2-(N-acyl)-aminobenzimidazoles. Tetrahedron Lett 2002;43:7303–6.

[9] Ries UJ, Mihm G, Narr B, Hasselbach KM, Wittneben H, Entzeroth M, van Meel JCA, Wienen W, Hauel NH. 6-Substituted benzimidazoles as new nonpeptide angiotensin II receptor antagonists: synthesis, biological activity, and structure-activity relationships. J Med Chem 1993;36:4040–51.

[10] Mahmoud Z, Daneshtalab M. Imidazoquinolines as diverse and interesting building blocks: review of synthetic methodologies. Heterocycles 2012;85:2651–83.

[11] Blaser HU. Reactions at surfaces: opportunities and pitfalls for the organic chemist; modern synthetic methods. John Wiley & Sons Inc; 1995. B. Ernst and C. Leumann p. 179.

[12] Power LA. PhD Thesis. University of St Andrews; 2008.

[13] Gebarowski P, Sas W. Asymmetric synthesis of novel polyhydroxylated derivatives of indolizidine and quinolizidine by intramolecular 1,3-dipolar cycloaddition of N-(3-alkenyl)nitrones. Chem Commun 2001;10:915–16.

[14] Emmons WD. Oxidation reactions with pertrifluoroacetic acid. J Am Chem Soc 1953;75:4623–4.

[15] Mullen P, Miel H, McKervey MA. N-Boc 4-Nitropiperidine: preparation and conversion into a spiropiperidine analogue of the eastern part of maraviroc. Tetrahedron Lett 2010;51:1–3.

[16] Li H, Zhang S, Yu C, Song X, Wang W. Organocatalytic asymmetric synthesis of chiral fluorinated quaternary carbon containing β-ketoesters. Chem Commun 2009;16:2136–2138.

[17] Ji J, Barnes DM, Zhang J, King SA, Wittenberger SJ, Morton HE. Catalytic enantioselective conjugate addition of 1,3-dicarbonyl compounds to nitroalkenes. J Am Chem Soc 1999;121:10215–16.

[18] Barnes DM, Ji J, Fickes MG, Fitzgerald MA, King SA, Morton HE, Plagge FA, Preskill M, Wagaw SH, Wittenberger SJ, Zhang J. Development of a catalytic enantioselective conjugate addition of 1,3-dicarbonyl compounds to nitroalkenes for the synthesis

of endothelin-A antagonistabt-546. Scope, mechanism, and further application to the synthesis of the antidepressant rolipram. J Am Chem Soc 2002;124:13097–105.
[19] Cue BW, Zhang J. Green process chemistry in the pharmaceutical industry. Green Chem Lett Rev 2009;2:193–211.
[20] Nichols PJ, DeMattei JA, Barnett BR, LeFur NA, Chuang T-H, Piscopo AD, Koch K. Preparation of pyrrolidine-based PDE4 inhibitors via enantioselective conjugate addition of α-substituted malonates to aromatic nitroalkenes. Org Lett 2006;8:1495–8.
[21] Cheruku SR, Padmanilayam MP, Vennerstrom JL. Synthesis of 2H-pyrroles by treatment of pyrrolidines with DDQ. Tetrahedron Lett 2003;44:3701–3.
[22] Tsuritani N, Yamada K, Yoshikawa N, Shibasaki M. Catalytic asymmetric syntheses of ICI-199441 and CP-99994 using nitro-Mannich reaction. Chem Lett 2002;31:276–7.
[23] Costello GF, James R, Shaw JS, Slater AM, Stutchbury NCJ. 2-(3,4-Dichlorophenyl)-N-methyl-N-[2-(1-pyrrolidinyl)-1-substituted-ethyl]acetamides: the use of conformational analysis in the development of a novel series of potent opioid kappa agonists. J Med Chem 1991;34:181–9.
[24] Hanessian S, Pham V. Catalytic asymmetric conjugate addition of nitroalkanes to cycloalkenones. Org Lett 2000;2:2975–8.
[25] Jen WS, Wiener JJM, MacMillan DWC. New strategies for organic catalysis: the first enantioselective organocatalytic 1,3-dipolar cycloaddition. J Am Chem Soc 2000;122:9874–5.
[26] Paras NA, MacMillan DWC. The enantioselective organocatalytic 1,4-addition of electron-rich benzenes to α,β-unsaturated aldehydes. J Am Chem Soc 2002;124:7894–5.
[27] Lubkoll J, Wennemers H. Mimicry of polyketide synthases -enantioselective 1,4-addition reactions of malonic acid half-thioesters to nitroolefins. Angew Chem, Int Ed 2007;46:6841–4.
[28] Bae HY, Some S, Lee JH, Kim J-Y, Song MJ, Lee S, Zhang YJ, Song CE. Organocatalytic enantioselective Michael-addition of malonic acid half-thioesters to β-nitroolefins: from mimicry of polyketide synthases to scalable synthesis of γ-amino acids. Adv Synth Catal 2011;353:3196–202.
[29] Ricci A. Asymmetric organocatalysis at the service of medicinal chemistry. ISRN Org Chem 2014:1–29.
[30] O'Neill BT, Thadeio PF, Bundesmann MW, Elder AM, McLean S, Bryce DK. Tandem Michael addition and azanorbornane substance - P antagonists. Tetrahedron 1997;53:11121–40.
[31] O'Connell KMG, Diaz-Gavilan M, Galloway WRJD, Spring DR. Two-directional synthesis as a tool for diversity-oriented synthesis: synthesis of alkaloid scaffolds. Beilstein J Org Chem 2012;8:850–60.
[32] Denmark SE, Schnute ME, Marcin LR, Thorarensen A. Nitroalkene inter [4+2]/intra [3+2] tandem cycloadditions. 7. Application of (R)-(-)-2,2-diphenylcyclopentanol as the chiral auxiliary. J Org Chem 1995;60:3205–20.
[33] Denmark SE, Marcin LR. Asymmetric nitroalkene [4+2] cycloadditions: enantioselective synthesis of 3-substituted and 3,4-disubstituted pyrrolidines. J Org Chem 1995;60:3221–35.
[34] Denmark SE, Schnute ME. Tandem inter [4+2]/intra [3+2] cycloadditions. 3. The stereochemical influence of the Lewis acid. J Org Chem 1991;56:6738–9.
[35] Denmark SE, Thorarensen A, Middleton DS. Tandem [4+2]/[3+2] cycloadditions of nitroalkenes. 9. Synthesis of (-)-rosmarinecine. J Am Chem Soc 1996;118:8266–77.
[36] Denmark SE, Thorarensen A. The tandem cycloaddition chemistry of nitroalkenes. A novel synthesis of (-)-hastanecine. J Org Chem 1994;59:5672–80.
[37] Denmark SE, Seierstad ME. Tandem cycloaddition chemistry of nitroalkenes: probing the remarkable stereochemical influence of the Lewis acid. J Org Chem 1999;64:1610–19.

[38] Denmark SE, Senanayake CBW, Ho G-D. Tandem [4+2]/[3+2]-cycloadditions. 2. Asymmetric induction with a chiral vinyl ether. Tetrahedron 1990;46:4857–76.
[39] Schlaf M, Bosch M. Synthesis of allyl and alkyl vinyl ethers using an in situ prepared air-stable palladium catalyst. Efficient transfer vinylation of primary, secondary, and tertiary alcohols. J Org Chem 2003;68:5225–7.
[40] Seebach D, Leitz HF, Ehrig V. Michael-additionen von lithiumenolaten und schwefel-substituierten lithiumorganylen an nitroolefine. Chem Ber 1975;108:1924–45.
[41] Yanami T, Kato M, Yoshikoshi A. Fluoride-catalysed Michael addition of simple nitro-olefins to β-diketones. J Chem Soc, Chem Commun 1975;17:726–7.
[42] Cory RM, Anderson PC, McLaren FR and Yamamoto BR. 1981. Bicycloannulation with nitroethene and 1-nitropropene. A one-step synthesis of tricyclenone. 12: 73-74.
[43] Stalo ML, Burger A. New synthesis of myosmine. J Am Chem Soc 1957;79:154–6.
[44] Seeman JI. Effect of conformational change on reactivity in organic chemistry. Evaluations, applications, and extensions of Curtin–Hammett/Winstein–Holness kinetics. Chem Rev 1983;83:83–134.
[45] Chavdarian CG, Seeman JI, Wooten JB. Bridged nicotines. Synthesis of cis-2,3,3a,4,5,9b-hexahydro-1-methyl-1H-pyrrolo[2,3-f]quinoline. J Org Chem 1983;48: 492–494.
[46] Zuo Z, Ma D. Enantioselective total syntheses of communesins A and B. Angew Chem, Int Ed 2011;50:12008–11.
[47] Mršić N, Minnaard AJ, Feringa BL, de Vries JG. Iridium/monodentate phosphoramidite catalyzed asymmetric hydrogenation of N-aryl imines. Tetrahedron: Asymmetry 2009;131:8358–9.
[48] Shen CC, Guan GH, Young CT. New method of preparing p-nitroacetophenone from auto-oxidation of p-nitroethylbenzene. Acta Pharm Sinica 1958;6:210–12.
[49] Imanari M, Iwane H, Kujira K, Seto T. Production of indole. JP Patent 1986:61134370.
[50] Layek M, Lakshmi U, Kalita D, Barange DK, Islam A, Mukkanti K, Pal M. Pd/C-Mediated synthesis of indoles in water. Beilstein J Org Chem 2009;5:1–9.
[51] Batcho AD, Leimgruber W. Indoles from 2-methylnitrobenzenes by condensation with formamide acetals followed by reduction: 4-benzyloxyindole. Org Synth 1985;63:214–25.
[52] Bentley JM, Adams DR, Bebbington D, Benwell KR, Bickerdike MJ, Davidson JEP, Dawson CE, Dourish CT, Duncton MAJ, Gaur S, George AR, Giles PR, Hamlyn RJ, Kennett GA, Knight AR, Malcolm CS, Mansell HL, Misra A, Monck NJT, Pratt RM, Quirk K, Roffey JRA, Vickers SP, Cliffe IA. Indoline derivatives as 5-HT 2C receptor agonists. Bioorg Med Chem Lett 2004;14:2367–70.
[53] Garden SJ, Torres JC, Ferreira AA, Silva RB, Pinto AC. A modified Sandmeyer methodology and the synthesis of (±)-convolutamydine A. Tetrahedron Lett 1997;38:1501–4.
[54] Wojciechowski K, Makosza M. A facile synthesis of 3-sulfonyl-substituted indole derivatives. Synthesis 1986;8:651–3.
[55] Bernotas RC, Antane S, Shenoy R, Le V-D, Chen P, Harrison BL, Robichaud AJ, Zhang GM, Smith D, Schechter LE. 3-(Arylsulfonyl)-1-(azacyclyl)-1H-indoles are 5-HT6 receptor modulators. Bioorg Med Chem Lett 2010;20:1657–60.
[56] Wojciechowski K, Makosza M. Synthesis of 2-arylindoles via condensation of ortho-aminobenzyl sulfones with aromatic aldehydes. Bull Soc Chim Belg 1986;95:671–3.
[57] Makosza M, Glinka T, Kinowski J. Specific ortho orientation in the vicarious substitution of hydrogen in aromatic nitro compounds with carbanion of chloromethyl phenyl sulfone. Tetrahedron 1984;40:1863–8.
[58] Heffernan GD, Coghlan RD, Manas ES, McDevitt RE, Li Y, Mahaney PE, Robichaud AJ, Huselton C, Alfinito P, Bray JA, Cosmi SA, Johnston GH, Kenney T, Koury E, Winneker RC, Deecher DC, Trybulski EJ. Dual acting norepinephrine reuptake inhibitors and 5-HT2A receptor antagonists: identification, synthesis and activity of

novel 4-aminoethyl-3-(phenylsulfonyl)-1*H*-indoles. Bioorg Med Chem 2009;17:7802–15.
[59] Makosza M, Wojciechowski K. Nucleophilic substitution of hydrogen in arenes and heteroarenes. Top Heterocycl Chem 2014;37:51–106.
[60] Ellis GP. Synthesis of fused heterocycles: acetal or aldehyde and amine. The chemistry of heterocyclic compounds - a series of monographs. Taylor EC, editor. John Wiley & Sons; 1987. Chapter 2, ISBN 0-17 1-91 43 1-2, ISBN 13: 078-0-17 1-9 143 1-0.
[61] Stowell JG. Original synthesis of ellipticine and 9-methoxyellipticine through triazole. PhD Dissertation. Davis: University of California; 1981. p. 121–8.
[62] Miller RB, Stowell JG. Total synthesis of ellipticine and 9-methoxyellipticine via benzotriazole intermediates. J Org Chem 1983;48:886–8.
[63] Sakamoto T, Satoh C, Kondo Y, Yamanaka H. Condensed heteroaromatic ring systems. XXII. Simple and general synthesis of 1*H*-pyrrolopyridines. Heterocycles 1992;34:2379–84.
[64] Cupps TL, Wise DS, Townsend LB. Synthetic strategies for 2,4-dimethoxypyrrolo[3,2-*d*]pyrimidine. J Org Chem 1983;48:1060–4.
[65] Dugar S. Synthetic approaches to the 6*H*-pyrido[4,3-*b*] alkaloids. PhD Dissertation. Davis: University of California; 1984. p. 185–6.
[66] Morrison RT, Boyd RN. Organic chemistry. 5th Ed. Newton, MA: Allyn and Bacen Inc; 1987. p. 524–6.
[67] Balkau F, Elmes BC, Loder JW. Synthesis of ellipticine intermediates: 6-amino-, 6-hydroxy-, and 6-methoxy-5,8-dimethylisoquinoline. Aust J Chem 1969;22:2489–92.
[68] Dugar S. Synthetic approaches to the 6*H*-pyrido[4,3-*b*] alkaloids. PhD Dissertation. Davis: University of California; 1984. p. 141–85.
[69] Knudsen KR, Risgaard T, Nishiwaki N, Gothelf KV, Jørgensen KA. The first catalytic asymmetric aza-Henry reaction of nitronates with imines: a novel approach to optically active β-nitro-α-amino acid- and α,β-diamino acid derivatives. J Am Chem Soc 2001;123:5843–4.
[70] Duveau DY, Yasgar A, Wang Y, Hu X, Kouznetsova J, Brimacombe KR, Jadhav A, Simeonov A, Thomas CJ, Maloney DJ. Structure-activity relationship studies and biological characterization of human NAD^+-dependent 15-hydroxyprostaglandin dehydrogenase inhibitors. Bioorg Med Chem Lett 2014;24:630–5.
[71] Raghunath M, Viswanathan CL. Synthesis, characterization and evaluation of novel 6-substituted benzimidazole-2-carbamates as potential antimicrobial agents. Int J Pharm Pharm Sci 2014;6:372–5.
[72] Smith DM and Tennant G. Benzimidazoles and congeneric tricyclic compounds, Part 1. P.N. Preston (Ed.). John Wiley & Sons, ISBN 0-471-03792-3 (v. 1), ISBN 0-471-08189-2 (v. 2).
[73] Mann J, Baron A, Opeku-Boahen Y, Johansson E, Parkinson G, Kelland LR, Neidle S. A new class of symmetric bisbenzimidazole-based DNA minor groove-binding agents showing antitumor activity. J Med Chem 2001;44:138–44.
[74] Vora JJ, Trivedi KP, Kshatriya RS. An improved synthesis of 2-mercapto-5-difluoromethoxy-1*H*-benzimidazole: an important medicinal intermediate. Adv Appl Sci Res 2011;2:89–93.
[75] Wu CY, Sun CM. Soluble polymer-supported synthesis of 2-(arylamino)benzimidazoles. Tetrahedron Lett 2002;43:1529–33.
[76] Huang KT, Sun CM. Liquid-phase combinatorial synthesis of aminobenzimidazoles. Bioorg Med Chem Lett 2002;12:1001–3.
[77] Lee J, Doucette A, Wilson NS, Lord J. Solid phase synthesis of chiral 2-aminobenzimidazoles. Tetrahedron Lett 2001;42:2635–8.
[78] Krchnak V, Smith J, Vagner J. A solid phase traceless synthesis of 2-arylaminobenzimidazoles. Tetrahedron Lett 2001;42:1627–30.

[79] Katritzky AR, Rees CW. Comprehensive heterocyclic chemistry, 5. Oxford: Pergamon Press; 1984. p. 631–2.
[80] Sharma BR, Pujari HK. Heterocyclic systems containing bridgehead nitrogen atom. Part LXII. Synthesis of thiazolo[3,2-*a*]benzimidazol-3(2*H*)-ones. Indian J Chem 1988;27:121–7.
[81] Ronne E, Grivas S, Olsson K. Synthetic routes to the carcinogen IQ and related 3*H*-imidazo[4,5-*f*]quinolones. Acta Chem Scand 1994;48:823–30.
[82] Ziv J, Knapp S, Rosen JD. Convenient synthesis of the food mutagen 2-amino-3-methylimidazo[4,5-*f*]quinoline (IQ), and IQ-D3. Synth Commun 1988;18:973–80.
[83] Sukalovic V, Andric D, Roglic G, Kostic-Rajacic S, Soskic V. Electrostatic surface potential calculation on several new halogenated benzimidazole-like dopaminergic ligands. Arch Pharm (Weinheim, Ger) 2004;337:376–82.
[84] Katritzky AR, Belyakov SA, Tymoshenko DO. Preparation of polymer-bound 1*H*-benzotriazole, a new potential scaffold for the compilation of organic molecule libraries. J Comb Chem 1999;1:173–6.

CHAPTER 2

Raney nickel-assisted nitro group reduction for the synthesis of *N*-, *O*-, and *S*-heterocycles

2.1 Introduction

Heterocycles make up a useful class of compounds because of their extensive variety of uses. They are major among all kinds of agrochemicals, pharmaceuticals, and veterinary products. The *N*-, *O*-, and *S*-heterocycles are being studied for a long time due to their biological properties. The literature is enriched with advanced findings about the formation and pharmacological activities of fused heterocycles. The benzimidazoles condensed with other heterocyclic compounds exhibit a wide range of biological activities [1–10].

The *N*-containing heterocyclic compounds have remarkable uses in material and pharmaceutical sciences. Their synthesis apparently needs an incorporation of *N*-containing functionality via C–C and C–N bond formation reactions. Because of their rich chemical reactivity, NO_2 compounds acts as beneficial building blocks for the formation of nitrogen-containing heterocyclic compounds [11].

Sabatier and Senderens [12] reported heterogeneous Ni catalyst as early as 1897 for gas-phase hydrogenation, since then it has been utilized for a broad range of reactions such as methanation, hydrogenolysis, steam reforming, hydrogenation, reductive amination, and desulfurization, either on an industrial scale or in the laboratory. In this chapter, Raney nickel catalysts and reactions catalyzed by them are explained from a synthetic organic chemistry perspective.

2.2 Reduction of NO_2 group for the synthesis of five-membered *O*- and *S*-heterocycles

A rare use of nitroalkenes was observed in a Baylis–Hillman-like reaction where a tertiary amine like DABCO (1,4-diazabicyclo[2.2.2]octane) was able to add to the unsaturated system to create a stabilized zwitterionic

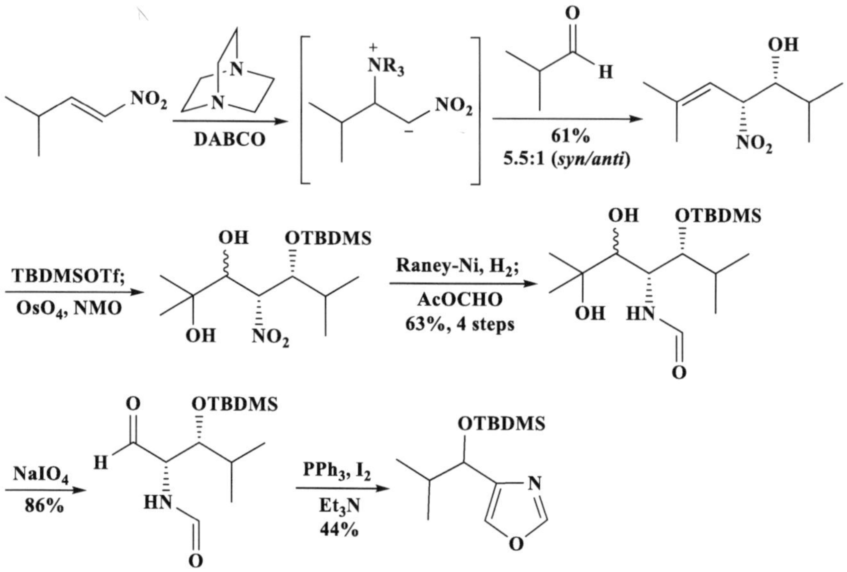

Scheme 2.1

system (Scheme 2.1) [13]. The reaction between this zwitterionic intermediate and isobutyraldehyde provided an unsaturated nitro alcohol with a certain preference for the *syn*-stereoisomer. The protection of OH group and osmylation of double bond furnished a compound, which upon reduction of NO₂ group followed by an oxidative cleavage of diol functionality provided amino aldehyde. The amino aldehyde could be transformed to a synthetic unit, oxazole, which is a part of various pharmacologically active medicines.

Stowell [14] and Miller et al. [15] prepared samples of intermediates required for elemental investigation to complete the set of experimental data for the formation of pyrazoloisoquinoline. A suitable precursor, 6-chloro-5,8-dimethyl-7-nitroisoquinoline, was accessible. It appeared to be very reluctant to react with thiophenylate in dimethylformamide, even though the bromo derivative was known to react well. The side-product 5,8-dimethyl-6,7-bis(thiophenyl)isoquinoline was observed after broad investigation. The consequent analysis about nucleophilic aromatic substitution showed [16] that the ease of displacement is in the sequence: fluoro > nitro > chloro, bromo, iodo. The desired 5,8-dimethyl-7-nitro-6-(thiophenyl)isoquinoline was invisible during this transformation as only thin layer chromatography was relied on to identify it, and it was identical to 6-chloro-5,8-dimethyl-7-nitroisoquinoline by thin layer chromatography. The product of primary reaction, after eliminating the side-product 5,8-dimethyl-6,7-bis(thiophenyl)isoquinoline, was the expected product 5,8-dimethyl-7-nitro-6-(thiophenyl)isoquinoline. The reduction of nitroisoquinoline with H_2NNH_2 and Raney-Ni in refluxing EtOH provided 7-amino-5,8-dimethyl-6-(thiophenyl)isoquinoline cleanly without the elimination of thiophenyl group. The diazotization of amine with $NaNO_2$ in cold HOAc provided a separable mixture of new ring system, 5-methyl-6-thiophenyl-3H-pyrazolo[3,4-h]isoquinoline and 5,11-dimethyl[1]benzothieno[2,3-g]isoquinoline (6-thiaellipticine). The vigor of nucleophilic displacement in dimethylformamide proposed that possibly the earlier required 7-azido-5,8-dimethyl-6-(thiophenyl)isoquinoline could be synthesized directly from nitroisoquinoline, and the reduction followed by diazotization of amine could be avoided, but do not avail. The nitroisoquinoline was reacted with azide in dimethylformamide to afford a complex mixture, may be because thiophenyl itself could be displaced by azide (Scheme 2.2).

2.3 Reduction of NO_2 group for the synthesis of six-membered N-heterocycles

The synthesis of piperidine nucleus needed an addition of homoenolate anion to nitroalkene for the preparation of 1,5-difunctional derivative. Since the accessibility of synthetic equivalents of homoenolate anions was not as wide as for enolates, the number of processes for the formation of piperidines using nitroalkenes was relatively less. The configurationally stable chiral azahomoenolate anions were available by lithiation of N-Boc allylic amines using (-)-sparteine (Scheme 2.3) [17,18]. The reaction of these anions with nitroalkenes afforded nitroenamines with satisfactory enantiomeric ratios and *syn*-selectivities. The hydrolysis of enamino groups and oxidation of intermediate aldehydes provided carboxylic acids, which on reduction of NO_2 groups and esterification afforded piperidin-2-one systems. Finally, the 3,4-disubstituted piperidines were prepared by the reduction of lactam groups and protection of the free secondary amines.

A successful synthetic method for both (+)- and (-)-pHTX was reported by Fitch and Luzzio [19,20] by an enzymatic desymmetrization and a double Henry reaction as the key steps in the formation of both antipodes of enantiomerically pure Kishi lactone (Scheme 2.4). Their pathway started with a double Henry condensation of nitroacetal and glutaraldehyde, which easily afforded crystalline nitrodiol in good yield. The triacetyl protection followed by a tandem ultrasound-assisted acetal deprotection/Wittig

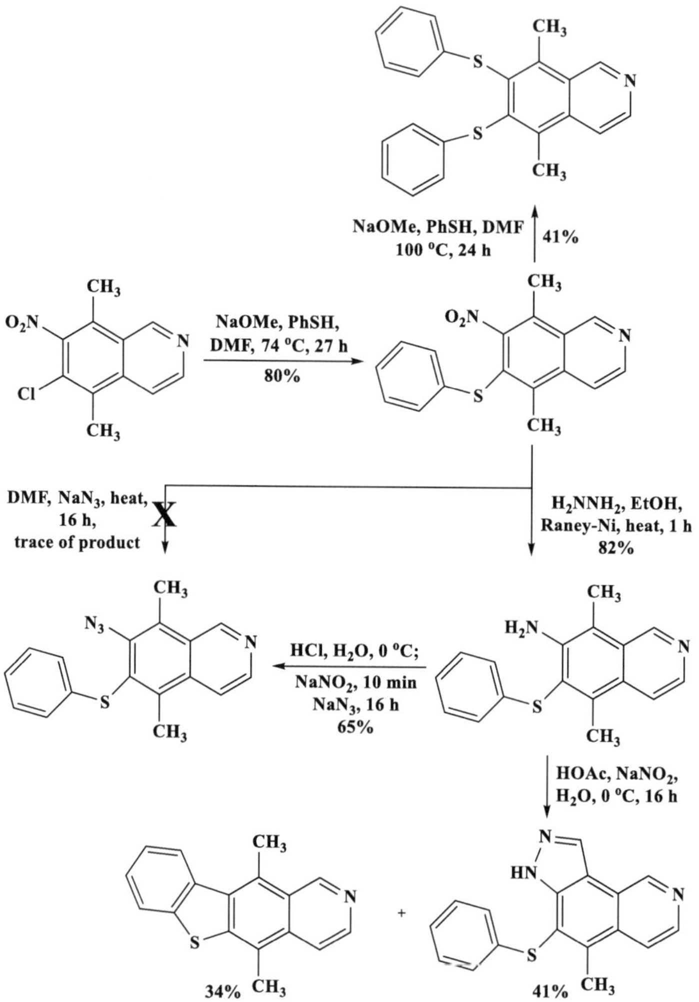

Scheme 2.2

homologation provided α,β-unsaturated ester in 70% yield. The saturated ester was obtained by catalytic hydrogenation of α,β-unsaturated ester. An acid-assisted hydrolysis of acetyl functions led to crude dihydroxyamino acid, which underwent an N,N'-dicyclohexylcarbodiimide-assisted lactamization to afford the core azaspirocyclic diol efficiently in excellent yield. Many enzymatic desymmetrizations were discovered for the formation of optically pure monoacetate. The first of these to be examined was the regioselective monoacylation of achiral spirodiol utilizing vinyl acetate and a

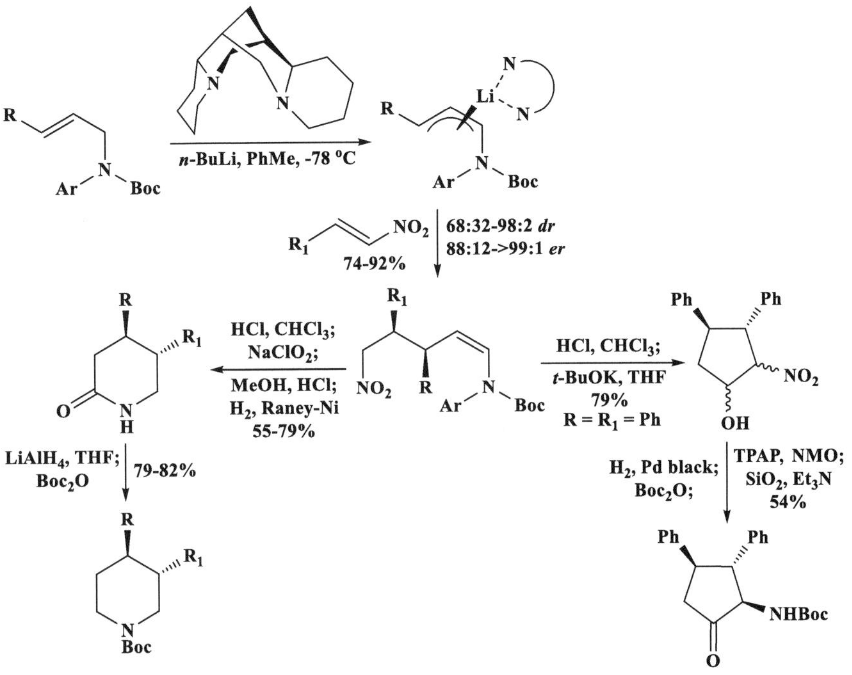

Scheme 2.3

variety of lipases and esterases. Disappointingly, all the attempted conditions were unsuccessful to afford any acylated products. Then, they synthesized *meso*-diacetylated compound by chemical approaches and further attempted an enzymatic monohydrolysis applying a variety of enzymes. The use of porcine liver esterase at pH 7.5 allowed them to isolate the crystalline (-)-monoacetate in 93% enantiomeric excess and 87% yield as determined by ^{19}F nuclear magnetic resonance examination of Mosher ester derivative. They also performed this enzymatic hydrolysis on triacetate, but observed this to be completely resistant to all the applied conditions. The enantiomerically pure alcohol was transformed to its phenylthiocarbonates and methylcarbonate, with the objective of carrying out a free-radical deoxygenation. The reaction of methylcarbonate and tri-*n*-butyltin hydride with AIBN only gives rise to complete recovery of the alcohol. The dehydroxy acetate was obtained in an excellent yield (93%) by treating phenylthiocarbonate under same reaction conditions. The enantiomerically pure Kishi lactam [20] was formed in 75% yield by deacylation followed by a Moffat oxidation [7].

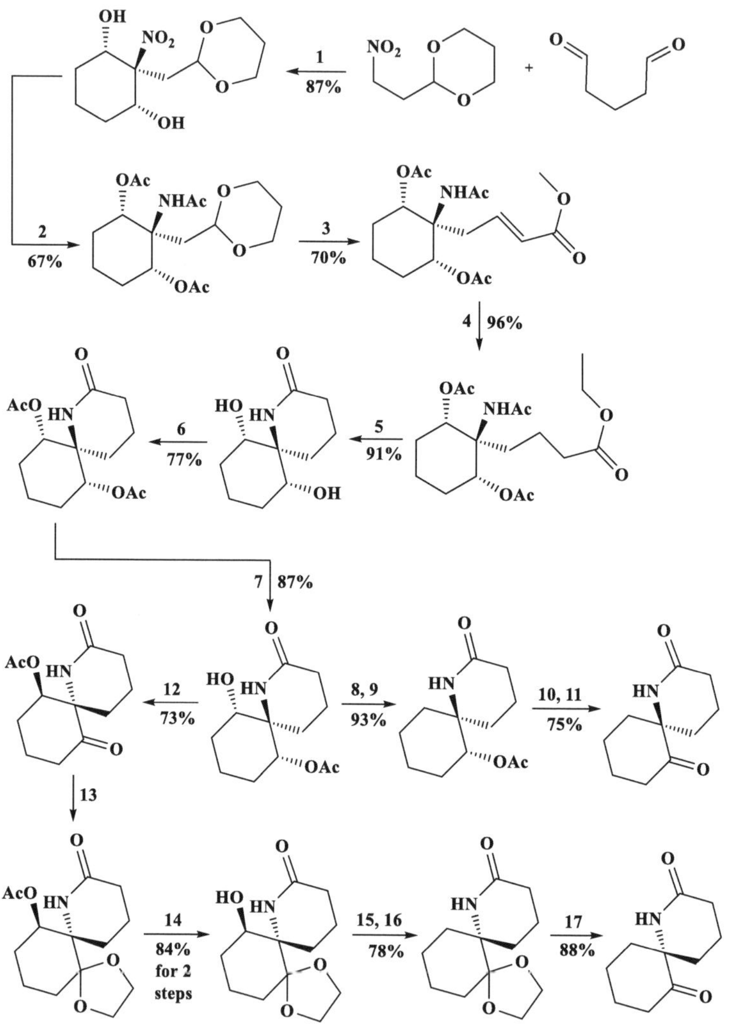

Reagents and conditions: (1) TMG, THF, (2) Al-Hg, THF, Raney-Ni, H$_2$; Et$_3$N, ultrasound, Ac$_2$O, Py, (3) AcOH, H$_2$O, ultrasound; EtOCOCH$_2$PPh$_3$$^+Br^-$, Et$_3$N, (4) H$_2$, Pd/C, MeOH, (5) 1.2 M HCl, reflux; DCC, DMAP, Py, (6) AcCl, DMAP, DCM, (7) pig liver esterase, pH 7, (8) PhOCSCl, DMAP, DCM, (9) Bu$_3$SnH, AIBN, toluene, 95 °C, (10) NaOMe, MeOH, (11) (COCl)$_2$, DMSO, DCM, (12) DMSO, DCC, H$_3$PO$_4$, (13) HOCH$_2$CH$_2$OH, TsOH, toluene, (14) NaOMe, MeOH, (15) PhOCSCl, DMAP, DCM, (16) Bu$_3$SnH, AIBN, toluene, 95 °C, (17) AcOH, TFA, H$_2$O.

Scheme 2.4

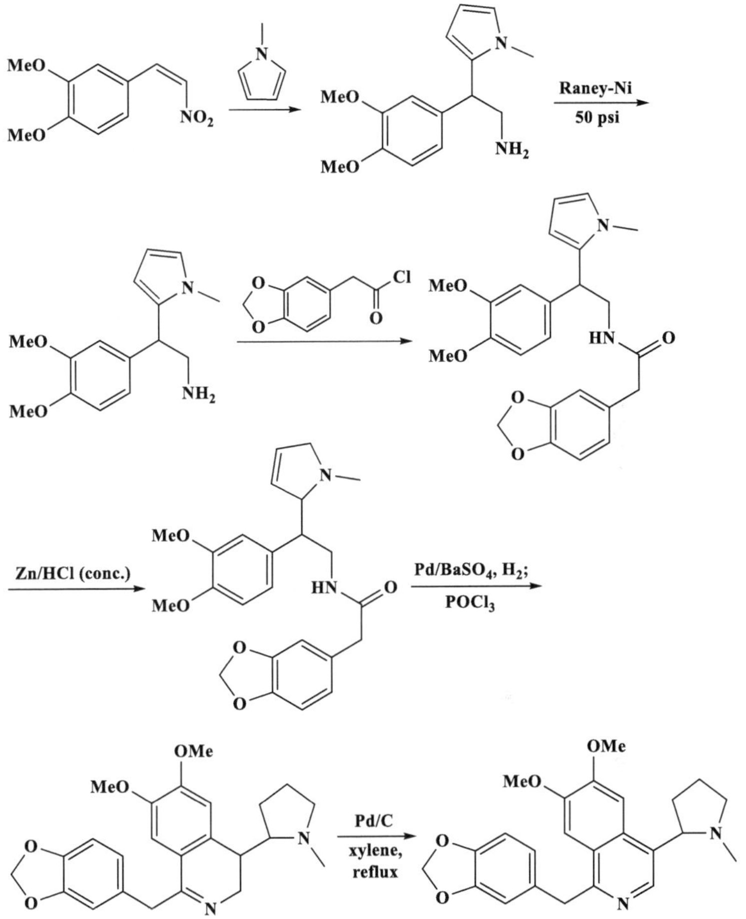

Scheme 2.5

The starting compound nitrostyrene was treated with *N*-methylpyrrole. The catalytic hydrogenation of formed compound and subsequent condensation with chloro compound afforded amide. The partial reduction of pyrrole ring led to dihydro intermediate. The pyrrole system was completely reduced to pyrrolidine intermediate by hydrogenation of dihydro intermediate with Pd. The reaction of reduced intermediate with POCl$_3$ afforded isoquinoline, which was aromatized with palladium/carbon in refluxing xylene to provide the (±)-target compound in 7% overall yield (Scheme 2.5) [21].

A novel formation of tetrahydroquinolines using 2-(2-nitroaryl)-3-aroyl-benzo[*b*]thiophene as starting compound was reported by Chabert

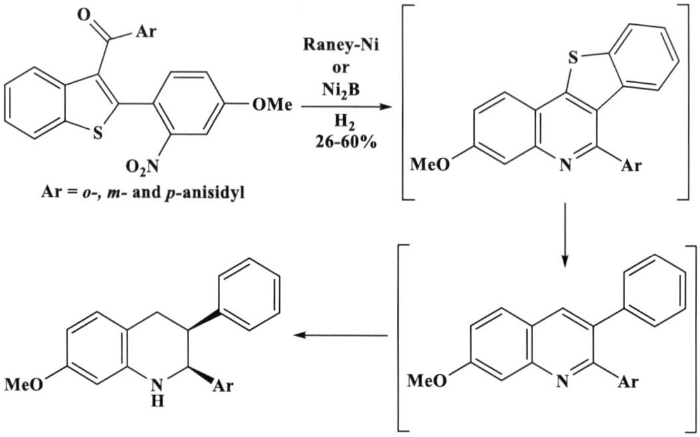

Scheme 2.6

et al. [22] by a Ni-catalyzed reduction-cyclization-desulfurization sequence. The reaction of benzo[*b*]thiophene derivatives with H$_2$ using Raney-Ni or nickel boride provided moderate to good yields of 2,3-diaryltetrahydroquinolines (Scheme 2.6) [8].

A rapid and high yielding CDC (cross-dehydrogenative coupling) reaction of CH$_3$NO$_2$ and *N*-phenyl tetrahydroisoquinoline provided nitroamine (Scheme 2.7). This oxidative coupling reaction was performed with variations by others, but never with any other protecting group on the nitrogen atom; this was unfortunate since synthetic elaboration of products led to a variety of versatile new chiral *vic*-diamines, analogous to those available with Reissert chemistry but without the hazards associated with the utilization of cyanide. An access to these types of compounds has been attained by aza-Henry reaction of dihydroisoquinoline. While the reduction of β-nitroamine to protected diamine was possible, the products could not be taken further. The phenyl substituent promoted the CDC reaction but essentially served as a protecting group that was very efficient in subsequent steps. These difficulties were overcome by utilizing an alternative PMP-protected starting compound derived from tetrahydroisoquinoline by a Cu-catalyzed arylation. The nitroamine was obtained by a cross-dehydrogenative coupling reaction with CH$_3$NO$_2$, which proceeded with similar efficiency and speed as the original reaction. The reduction of NO$_2$ group to amine, not a trivial process for β-nitroamines was effected with Ra-Ni in methanolic NH$_3$ at rt in excellent yield. An access to more interesting and diverse derivatives needed the removal of protecting group.

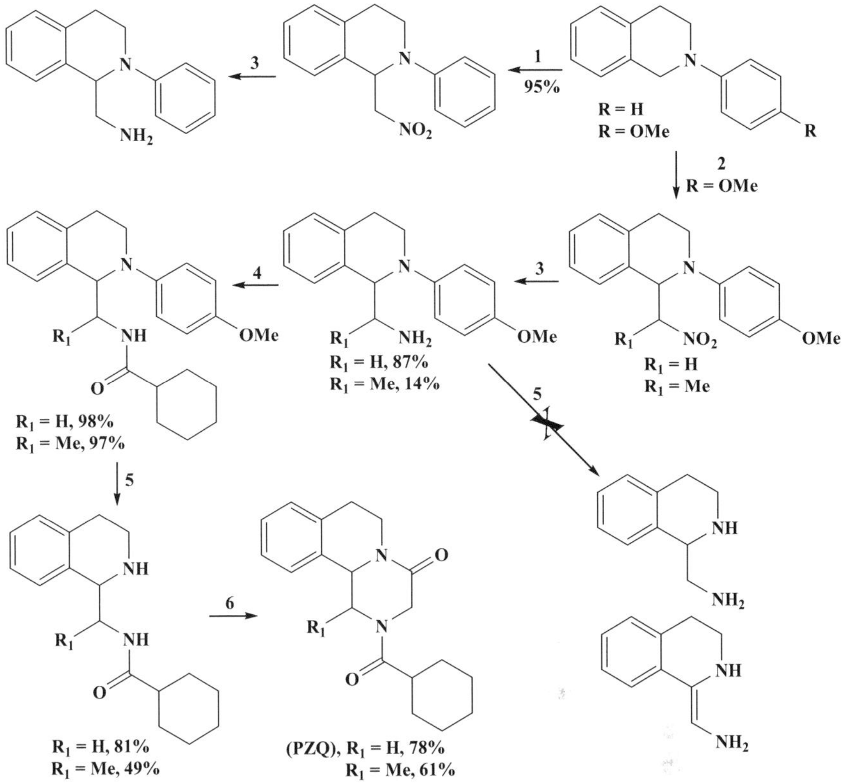

Scheme 2.7

Reagents and conditions: (1) iodoanisole, CuI, K$_3$PO$_4$, ethylene glycol, isopropanol, 80 °C, 24 h, (2) DDQ, nitromethane for a or nitroethane for b, rt, 5 min, (3) Raney-Ni, H$_2$, 4 h, (4) cyclohexane carbonyl chloride, DMAP, Et$_3$N, CH$_2$Cl$_2$, 0 °C, 4 h, (5) (NH$_4$)$_2$Ce(NO$_3$)$_6$, MeCN-H$_2$O, 0 °C, 5 min, (6) chloroacetyl chloride, NaOH, CH$_2$Cl$_2$, rt, 30 min, then TEBAC, reflux, 2 h.

It was expected that an oxidative deprotection of amine to diamine would be difficult because of the reactivity of amine. The enamine was obtained when CAN (ceric ammonium nitrate, (NH$_4$)$_2$Ce(NO$_3$)$_6$) was used. The amine was transformed to an amide to moderate the reactivity of amine; with one eye on an eventual formation of PZQ by reacting amine with cyclohexane carbonyl chloride to provide the amide. Literature processes explaining the elimination of *p*-methoxyphenyl groups from nitrogen centers under oxidative conditions used wet solvents for the hydrolysis of electron-rich ring from the heteroatom, providing free amides or amines and the benzoquinone fragment. It was observed that a small excess of ceric ammonium nitrate in aqueous CH$_3$CN efficiently eliminate the PMP

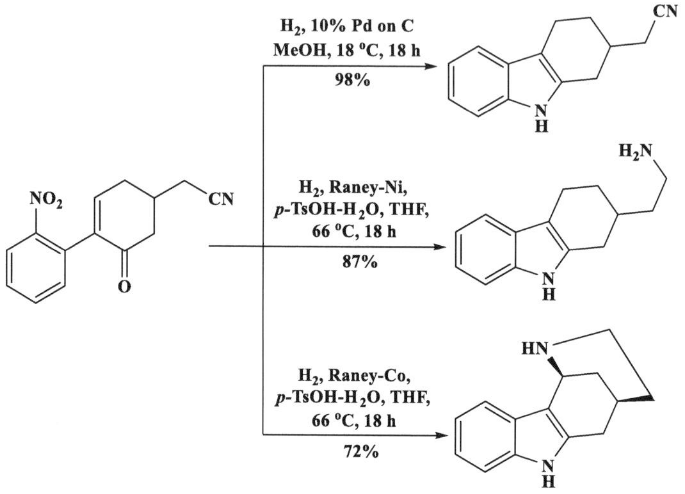

Scheme 2.8

group, affording free amine in good yield. The subsequent cyclization with ClCH$_2$COCl under phase-transfer conditions was also successful and gave praziquantel in 45% overall yield [23].

An easily available and polyfunctionalized cyclohexenone under a variety of reductive cyclization conditions afforded either partially cyclized and partially reduced, partially cyclized and fully reduced or completely cyclized and fully reduced products (Scheme 2.8) [24]. Then, the reaction with dihydrogen using 10% palladium on carbon in MeOH at 18°C provided indole (98%) as a result of reduction of the carbon–carbon bond and reductive cyclization of the NO$_2$ group on the pendant carbon–oxygen (no specific order of events implied). Under these conditions, the associated nitrile residue was not clearly affected. The *p*-toluenesulfonic acid monohydrate, obtained on exposure to dihydrogen using Raney-Ni and *p*-TsOH·H$_2$O, was introduced to avoid the reductive alkylation reactions including intermediates obtained by reduction of nitrile group. The cyclohexenone was exposed to reactions conditions but, additionally, the nitrile group was also reduced to primary amine (and so providing amine compound in 87% yield). The amine was also formed in 98% yield when indole was reacted with dihydrogen and Raney-Ni. Finally, and most importantly, two reductive cyclization events occurred when cyclohexenone was reacted with dihydrogen, Raney-Co and *p*-toluenesulfonic acid monohydrate, one event involved an intramolecular hetero-Michael addition reaction (providing a piperidine ring) and the other event was an intramolecular Schiff base-type

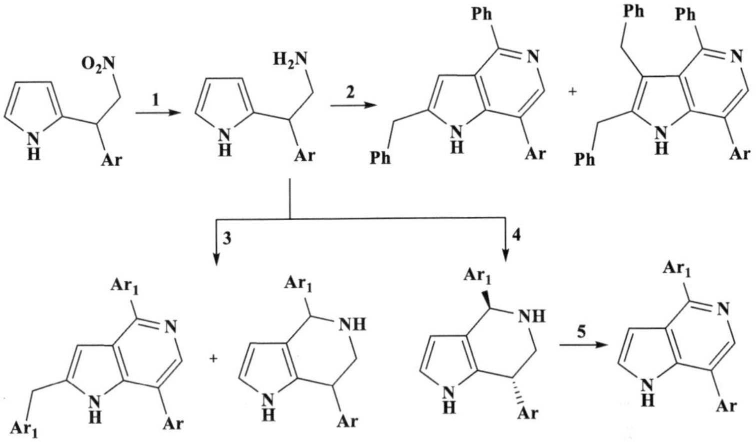

Reagents and conditions: (1) Raney-Ni, H$_2$, MeOH, (2) PhCHO, DCM, Si(CH$_3$)$_3$Cl, 0 °C to rt, 48 h, (3) Ar$_1$CHO, DCM, TFA, rt, 12-24 h, (4) Ar$_1$CHO, DCM, AcOH (glacial), rt, 36-38 h, (5) 5% Pd/C, xylene, reflux, 2-15 h.

Scheme 2.9

condensation reaction (affording an indole functionality). Subsequently, the tetracyclic compound was formed in 72% yield. The transformation cyclohexenone → tetracyclic compound was of specific interest because the product represented the 1,5-methanoazocino[4,3-b]indole framework associated to *Strychnos* and *Uleine* alkaloids.

Michael addition of pyrrole on nitro olefins gave nitro adducts, which were reduced to not reported 2-(2-pyrrolyl)-1-aminoethanes. The Pictet-Spengler condensation utilizing glacial CH$_3$COOH provided novel 4,7-disubstituted 4,5,6,7-tetrahydro-5-azaindoles in diastereoselective manner (Scheme 2.9). The stereochemistry was examined by single crystal X-ray study. It showed *trans*-geometry as *R,S*-configuration at C-4 and C-7, respectively. The dehydrogenation of tetrahydro-5-azaindoles gave novel 4,7-disubstituted 5-azaindoles. The Pictet-Spengler condensation utilizing trimethylsilyl chloride or trifluoroacetic acid afforded novel substituted 5-azaindoles directly [24].

Ng et al. [25] reported a diastereoselective cycloaddition reaction between imines and arylthio-substituted succinic anhydrides to synthesize the γ-lactams in high yields and exploited the established process for the formation of a pyrroloquinoline skeleton of martinelline alkaloids. The lactam, formed via a cycloaddition-based process, underwent an intramolecular reductive cyclization with 1,8-diazabicyclo[5.4.0]undec-7-ene and Raney-Ni with the concomitant removal of S-(p-tolyl) group to

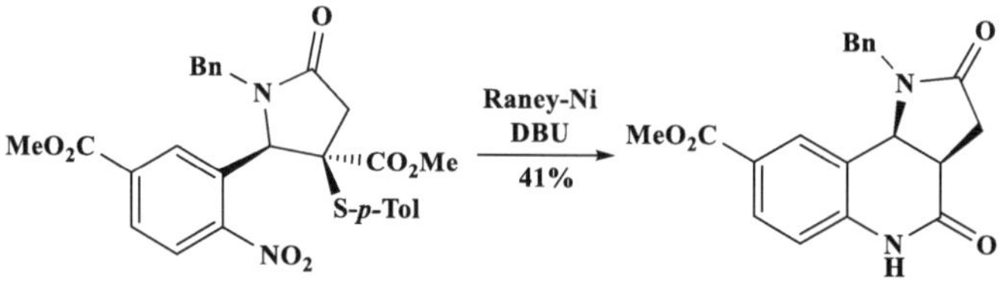

Scheme 2.10

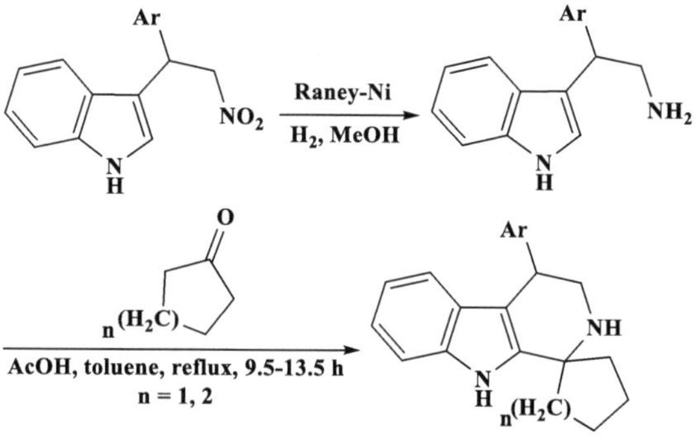

Scheme 2.11

afford the pyrroloquinoline in 41% yield (Scheme 2.10). A rather similar cyclization was earlier reported in high yields using sodium hydrosulfite beginning from substrates related to lactam but possessing no aryl thio group [8, 26,27].

The THSβCs (tetrahydrospiro-β-carbolines) are useful indole alkaloids as they exhibit many biological activities such as antispasmodic action, antidepressant activity, GHSR inhibitory action with Ki = 60 nM, anticonvulsant activity, and receptor affinity. Having Michael adducts in hand, the NO_2 compounds were reduced to amino compounds. Further, the reaction with ketones and Pictet-Spengler reaction in the presence of glacial AcOH afforded only one diastereomer of unreported 1,1,4-trisubstituted 2,3,4,9-tetrahydrospiro-β-carbolines (Scheme 2.11). The stereoselectivity in the reaction was demonstrated utilizing unsymmetrical ketones and single crystal X-ray study of spiro products, which showed the formation of only (R,R)-diastereomer [8].

The 3,6-diazacarbazole is a biologically active compound. A series of diazacarbazoles show various biological activities; therefore, these compounds are of interest as structural units with widespread applications in drug design. With Michael adduct of pyrrole in hand, it was decided to

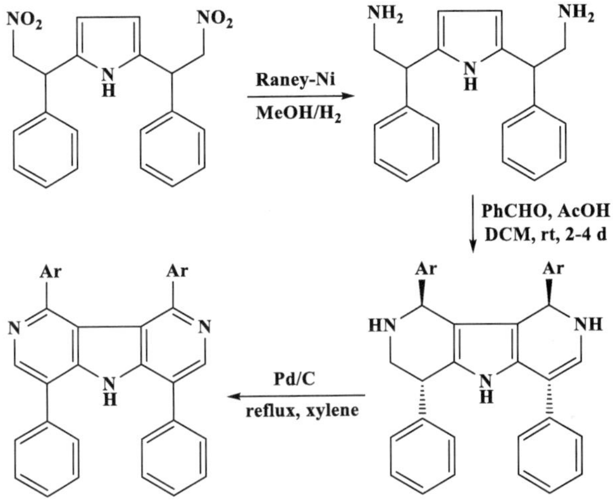

Scheme 2.12

reduce the NO$_2$ compounds to amino compounds. Then, the reaction with aldehyde and Pictet-Spengler reaction utilizing glacial AcOH provided only one diastereomer of unreported tetrasubstituted 3,6-diazacarbazole and tetrasubstituted octahydro-3,6-diazacarbazole (Scheme 2.12). The stereoselectivity in the reaction was demonstrated utilizing aldehydes bearing electron-withdrawing and electron-releasing groups and single crystal X-ray study of tetrasubstituted octahydro-3,6-diazacarbazole [8].

2.4 Reduction of NO$_2$ group for the synthesis of six-membered *N,N*-heterocycles

The tetraarylarenes were prepared in below 3% yield by coupling reactions utilizing crude 1,3,6,8-tetrabromo-4,5,9,10-tetraazapyrene. Therefore, an alternative technique was reported for the formation of these derivatives (Scheme 2.13), depending on the reported synthesis of the parent tetraazapyrene [28]. The tetraarylbiphenyl, obtained by Suzuki coupling reaction of suitable boronic ester and tetrachloride derivative, was smoothly purified by column chromatography. The tetraazapyrenes were obtained when two azo bridges were closed by the reduction of tetraarylbiphenyl with Ra-Ni under basic conditions. The procedure reported in literature was modified utilizing PrOH in place of EtOH to increase the solubility of biphenyls. The total yields (20–29%) for tetraazapyrenes were greater than those in the first technique. An overall synthesis improved considerably in the second

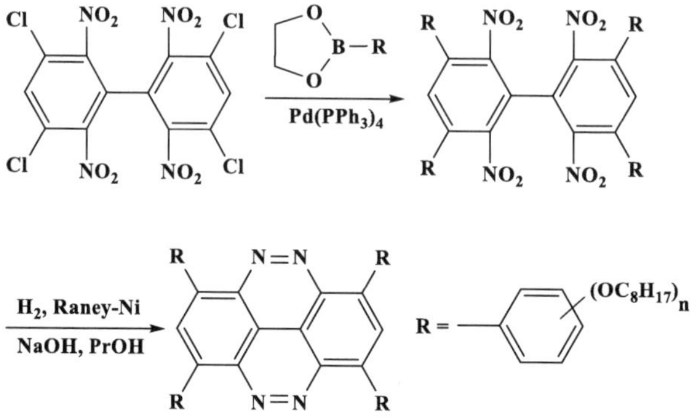

Scheme 2.13

technique by the ease of the reaction monitoring, purification of products and their intermediates, and generally greater yields [29].

The formation of pyrazolo[4,3-*d*]pyrimidin-7-ones was started with the synthesis of pyrazole ring from hydrazine hydrate and diketoester (Scheme 2.14). The substituted pyrazole intermediate was generated by regioselective *N*-alkylation of pyrazole ring, followed by nitration, carboxamide formation, and NO_2 group reduction. Under basic conditions, the pyrazolopyrimidinones were prepared by an acylation of amine with 2,4-substituted benzoyl chloride and cyclization. For 2-alkoxy series, the chlorosulfonylation occurred selectively at the 5-position of the phenyl ring which allowed ready coupling with a series of amines to provide the sulfonamide products. The synthesis of pyrazolopyrimidinones as described in the literature [30] involved hydrolization to afford the carboxylic acid and then nitration of these compounds with a mixture of fuming HNO_3 and oleum. The required amides were prepared from acid by chlorination with $SOCl_2$ followed by work-up in NH_4OH. In a modified methodology, the reaction occurred in two steps including nitration and aminolysis, and the synthetic pathway was shortened and the yields increased. Then, the catalytic reduction to amines was attempted using hydrazine hydrate and Ra-Ni. This reduction technique afforded pure compounds in good yields and utilized in next step without isolation and purification [31].

The *Cinchona* alkaloid-based catalysts represent a class of privileged organocatalysts for the promotion of conjugate addition reactions, involving those producing quaternary carbons. These are powerful bifunctional catalysts; usually the quinuclidine functionality in the catalyst activated

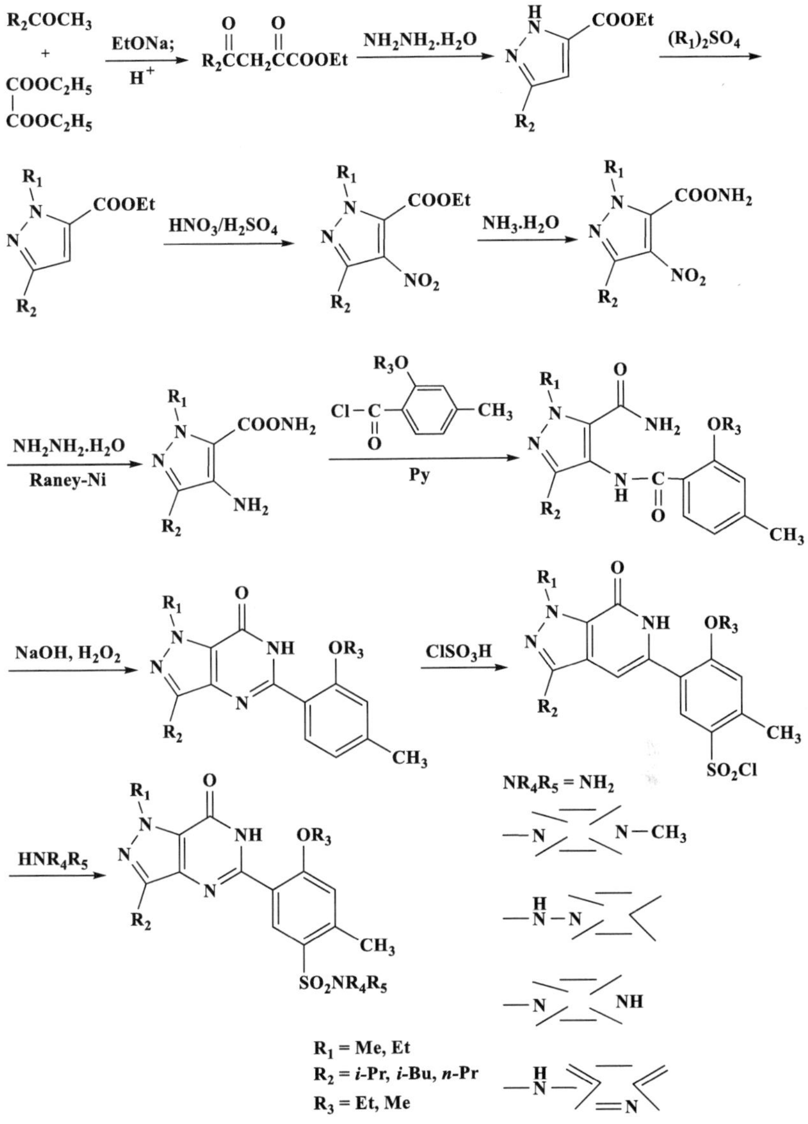

Scheme 2.14

the nucleophile, while the functional group thiourea or hydroxy at C-60 or C-9 activated the electrophile, both through the hydrogen bonding interactions. A *Cinchona* alkaloid-catalyzed Michael addition reaction was reported by Zhu et al. [32–33]. In this reaction, many aryl-isocyanoacetates were added to vinyl phenylselenone under the action of 6'-hydroxy

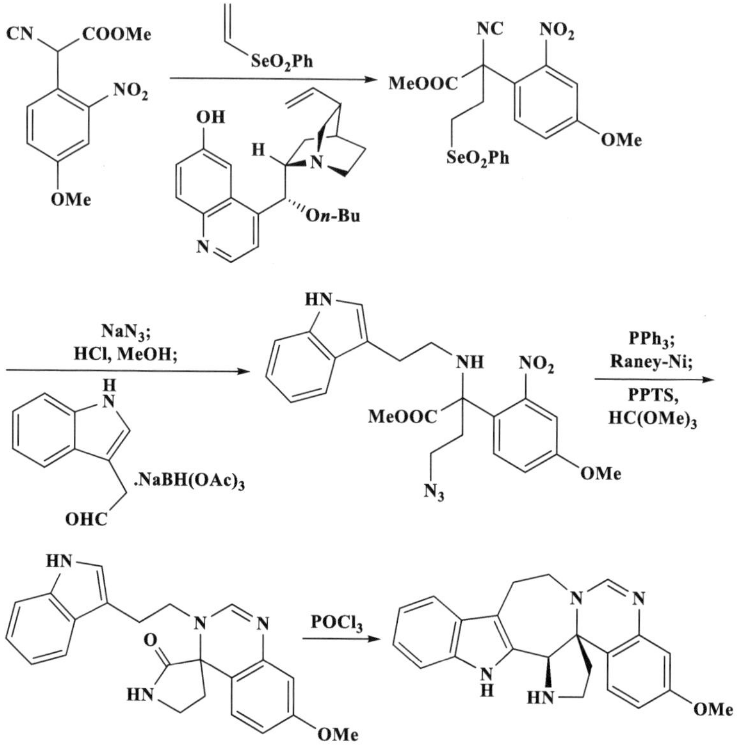

Scheme 2.15

Cinchona alkaloids to provide the synthetically beneficial isocyano-bonding quaternary carbons in good enantioselectivity and yields. They reported an overall formation of trigonoliimine A by this new approach (Scheme 2.15). The catalyzed Michael addition reaction of starting compounds provided products in 87% enantiomeric excess and 62% yield, which were transformed to indole derivative by a sequence including nucleophilic displacement of phenylselenone by NaN_3, acidic hydrolysis of isocyano group, and reductive amination of amine with aldehyde. The azide and NO_2 groups were reduced before being converted to azaspiro compound, which was reacted with phosphorus oxytrichloride to effect the Bischler–Napieralski reaction to give the trigonoliimine A [9,34].

The 2,5-dichloronitrobenzene was substituted with PMB (*p*-methoxybenzylamine) to afford the benzylamine, which was hydrogenated with Raney-Ni and H_2 in tetrahydrofuran to provide the aniline. The ring-closure of aniline to *N*-protected quinoxaline was performed by heating in pure diethyloxalate. The quinoxaline was chlorinated with phosphorus oxytrichloride using Hünig's base and dimethylformamide to synthesize the chloroquinoxaline. Similar to a process reported in the literature, an elimination of *p*-methoxybenzylamine group from chloroquinoxaline was achieved with conc. H_2SO_4 to provide the monochloro-substituted quinoxalinone [35]. Finally, the quinoxaline was formed regioselectively via this pathway by substitution of 2-Cl position of the monochloro-substituted quinoxalinone with *N*-methylpiperazine. A comparison of melting point, nuclear magnetic resonance, and liquid chromatography-mass spectrometry data revealed that the compounds obtained with both synthetic pathways were indeed identical. Similar process was utilized for the formation of quinoxaline from quinoxalines and quinoxalinone with Boc-protected 3-aminomethylpyrrolidine and *N*-methylhomopiperazine, respectively. The removal of Boc group with hydrochloric acid/dioxane finally provided hydrochloric salt of product (Scheme 2.16) [36].

The cyclization of α-amino acid intermediates, generated from *o*-nitrohalogenobenzene and amino acid, was an unambiguous technique for the formation of quinoxaline-2-ones (Scheme 2.17). The 2,4-dichloronitrobenzene provided 6-(*N*-2'-chloroethylamino)aniline, which was cyclized to 6-chloro-1,2,3,4-tetrahydroquinoxaline in 52% yield (Scheme 2.18) [37].

The ZK202000 was prepared by a stepwise regioselective nucleophilic displacement of the halogen atoms in 4,6-dichloronitro-3-(trifluoromethyl)-benzene. The Cl atom in the *ortho*-position to the NO_2 group was displaced by aminomethanephosphonic acid affording [[[5-chloro-2-nitro-4-(trifluoromethyl)phenyl]amino]methyl]phosphonic acid. Second, the ZK202000 was synthesized by preparing diethyl [[[5-chloro-2-nitro-4-(trifluoromethyl)phenyl]amino]methyl]phosphonate via esterification of acid with $HC(OEt)_3$ under acid catalysis. The nucleophilic displacement of second Cl atom by a styryl functionality was achieved by Pd catalyzed coupling to generate the (*E*)-diethyl [[[2-nitro-4-(trifluoromethyl)-5-(phenylethenyl)phenyl]amino]methyl]phosphonate. Simultaneous reduction of NO_2 group and styryl double bond was achieved by hydrogenation in EtOH utilizing Ra-Ni as a catalyst to afford the diethyl[[[2-amino-4-(trifluoromethyl)-5-(phenylethenyl)phenyl]amino]methyl]phosphonate. For the completion of the both syntheses, in a two-step sequence the quinoxalinedione scaffold was constructed by condensing phenylenediamine with ethyl oxalyl chloride followed by ring-closure upon heating the crude product in ethanolic hydrogen chloride. The ester

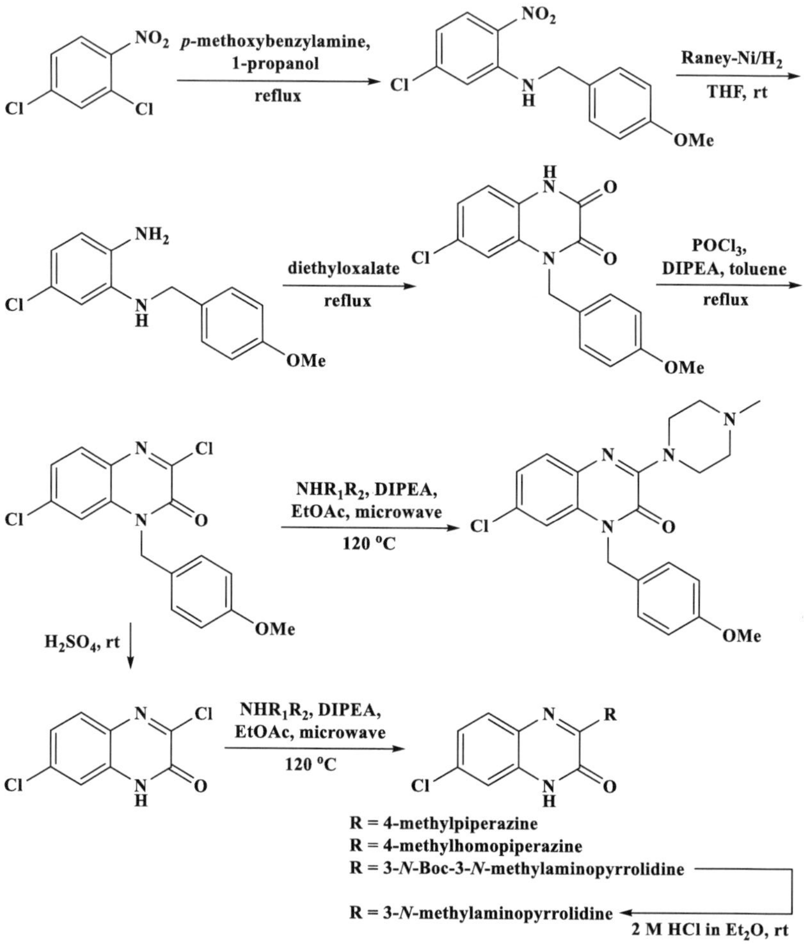

Scheme 2.16

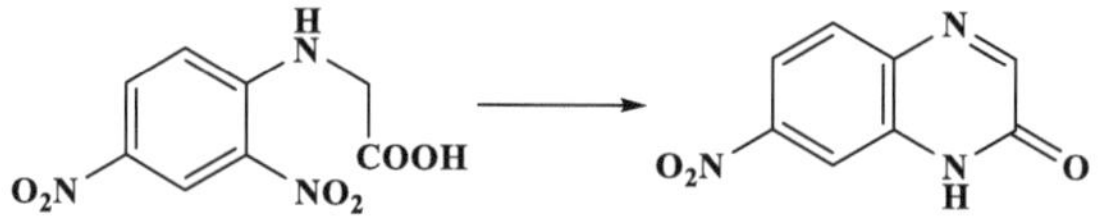

Scheme 2.17

cleavage with conc. aqueous HCl under heating led to [[1,2,3,4-tetrahydro-7-(phenylethyl)-2,3-dioxo-6-(trifluoromethyl)quinoxalin-1-yl]methyl]phosphonic acid (ZK 202000). The total yield of six-step reaction was >27% for ZK202000 (Scheme 2.19) [38].

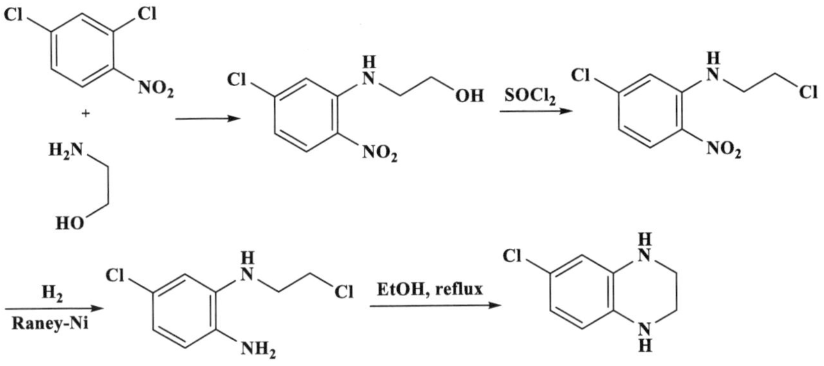

Scheme 2.18

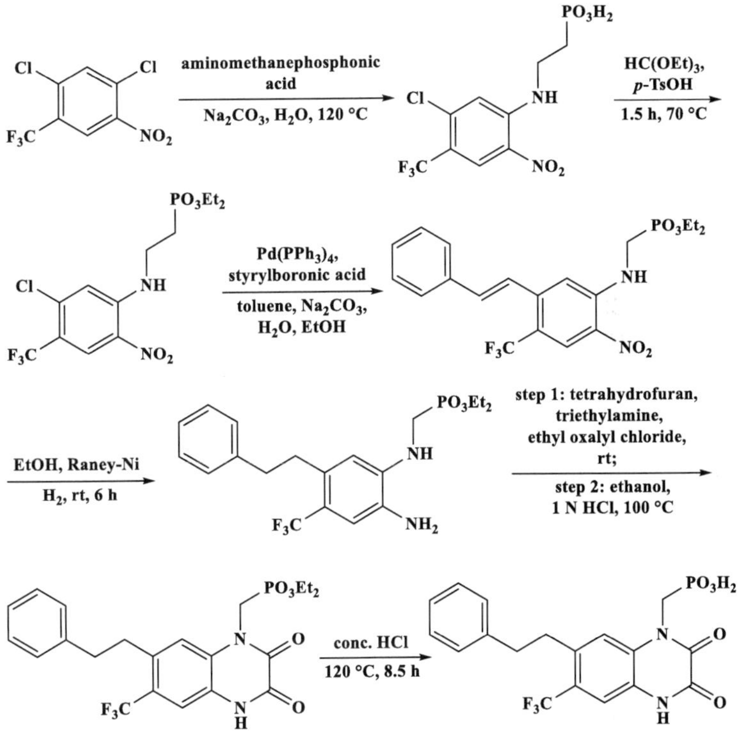

Scheme 2.19

The pyridopyrazine analogs were synthesized by the displacement of 4-Cl group of chloro compound with 2-amino-5-diethylaminopentane to give the diamine [39]. The reduction of 5-NO$_2$ group of diamine afforded triamine, and direct coupling with diverse diketones provided pyridopyrazines in good yields. Likewise, for the formation of other targets, the displacement of 4-Cl group of chloro compound with diverse amino compounds provided nitro amine compound, and the reduction of 5-NO$_2$ group with Raney-Ni and hydrogen at atmospheric pressure and rt provided diamine in quantitative yields. Final reaction with benzil, furil, and 2,2'-thenil delivered desired targets in good yields (Scheme 2.20) [40–42].

2.5 Reduction of NO$_2$ group for the synthesis of six-membered *N,N,N*-heterocycles

The 5-chloro-2-nitrobenzoyl chloride was reacted with methyl 4-amino-1-methyl-3-*n*-propylpyrazole-5-carboxylate to afford the crystalline compound, which was characterized as methyl 4-*N*-(5-chloro-2-nitrobenzoyl) amino-1-methyl-3-*n*-propyl-1*H*-pyrazole-5-carboxylate. The catalytic reduction of nitroaroylaminopyrazoles with Raney nickel easily gave methyl 4-*N*-(2-amino-5-chlorobenzoyl)amino-1-methyl-3-*n*-propyl-1*H*-pyrazole-5-carboxylate. The pyrazole ester was refluxed in NH$_3$ followed by work-up to provide the desired product, 4-*N*-(2-amino-5-chlorobenzoyl)amino-1-methyl-3-*n*-propyl-1*H*-pyrazole-5-carboxamide. The reaction of pyrazole diamides with NaNO$_2$ in HCl furnished tetracyclic derivative (Scheme 2.21) [10].

The linearly fused 11-methyl-9-*n*-propyl-5,6,11,12-tetrahydro-8*H*-benzo[*e*]pyrazolo[4',3':4,5][1,2,3]triazino[1,2-*a*][1,2,3]triazin-5,12-dione was prepared (Scheme 2.22). The 1-methyl-4-nitro-3-*n*-propylpyrazole-5-carboxylic acid was treated with SOCl$_2$ followed by condensation with 2-aminobenzamide to afford the 1-methyl-4-nitro-3-*n*-propylpyrazole-5-(*N*-2-carboxamidophenyl)carboxamide. The catalytic hydrogenation of 4-nitropyrazole-5-carboxamides with Raney nickel provided 4-aminopyrazolecarboxamides easily. The 11-methyl-9-*n*-propyl-5,6,11,12-tetrahydro-8*H*-benzo[*e*]pyrazolo[4',3':4,5][1,2,3]triazino[1,2-*a*][1,2,3]triazin-5,12-dione was obtained by the reaction of 4-aminopyrazole-5-carboxamides with NaNO$_2$ in HCl followed by usual work-up [10].

2.6 Reduction of NO$_2$ group for the synthesis of six-membered *O*-heterocycles

The second resolution method depends on the enzymatic resolution of acetate esters (Scheme 2.23) [43]. The sequence started with the alkylation of 2,3-difluoro-6-nitrophenol with 1-acetoxychloro-2-propane to provide

Raney nickel-assisted nitro group reduction for the synthesis of N-, O-, and S-heterocycles 63

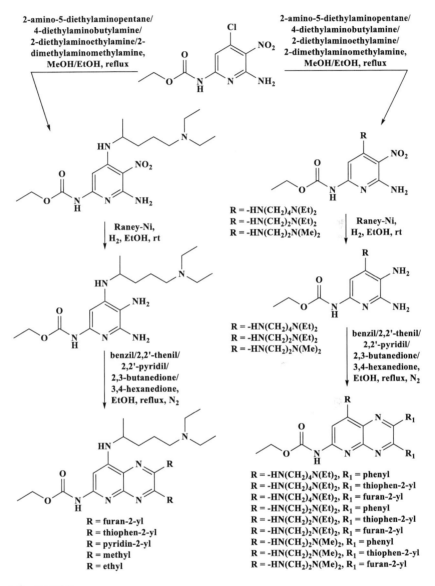

Scheme 2.20

the ether. The reduction of NO$_2$ group of ether provided an intermediate aniline, which was cyclized to provide the racemic benzoxazine in 62% yield. Then, many lipases were studied for the resolution. The best results came from utilization of LPL amano, derived from *P. aeruginosa*, which provided a ratio of 73:23. The enantiomerically pure final compound was

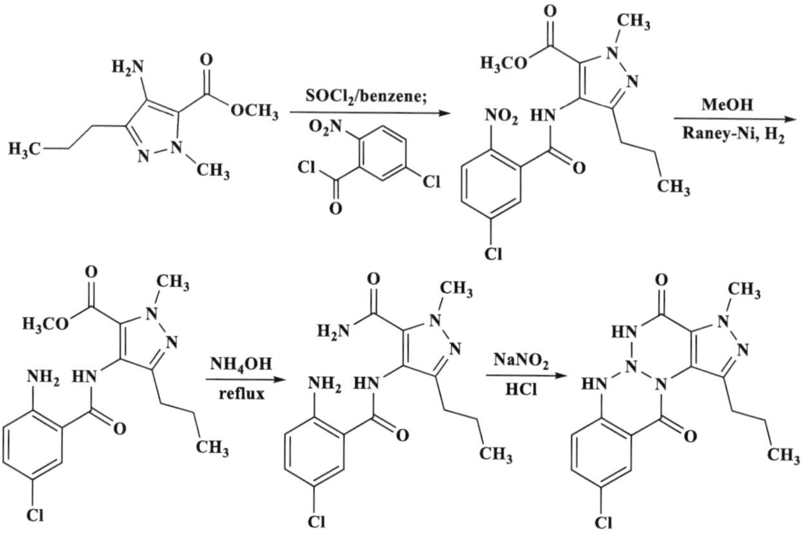

Scheme 2.21

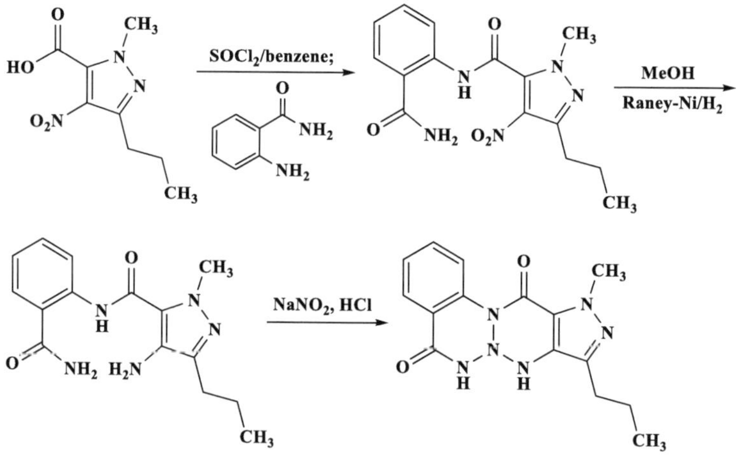

Scheme 2.22

obtained by benzoylation of enantiomerically enriched mixture followed by chromatography of aryl amides.

The benzoxazine substrate was synthesized by a linear sequence beginning from phenol, which was alkylated with epichlorohydrin (Scheme 2.24) [44]. The epoxide ring was opened with MeOH using Sn Lewis acid to afford the alcohol, which underwent Jones-oxidation and Ra-nickel-assisted hydrogenation to give the benzoxazine in moderate

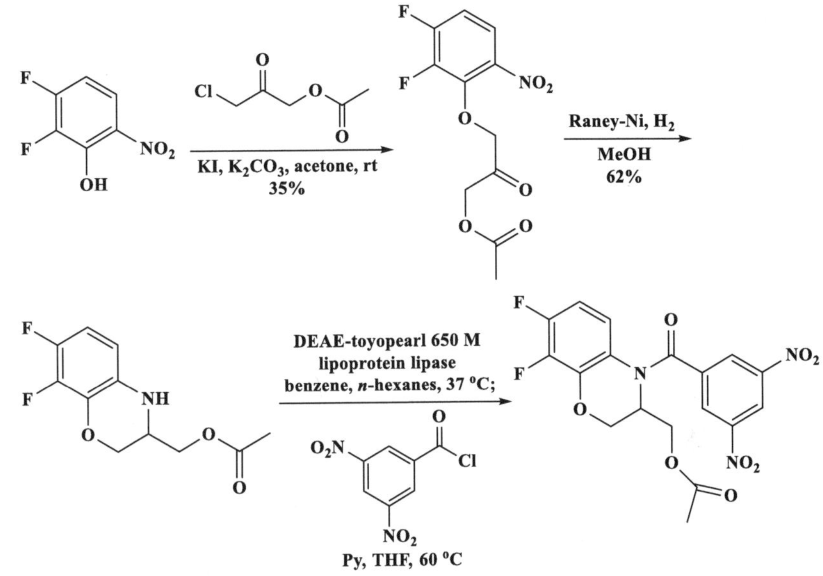

Raney nickel-assisted nitro group reduction for the synthesis of N-, O-, and S-heterocycles 65

Scheme 2.23

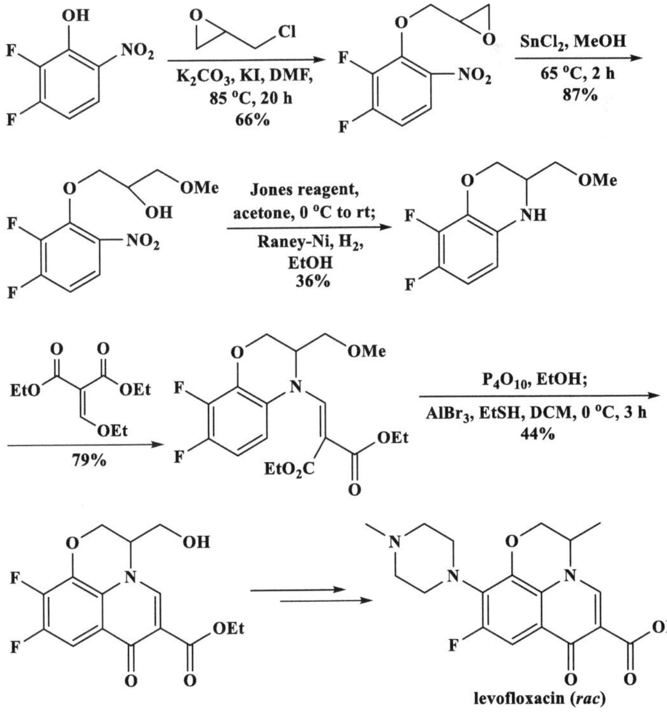

Scheme 2.24

yield through an imine intermediate. The imine intermediate was condensed with diethyl ethoxymethylenemalonate, and then cyclized by an intramolecular Friedel-Crafts acylation promoted by polyphosphate. The AlBr$_3$-mediated demethylation of pendant OMe group of this tricyclic structure provided alcohol, which was smoothly transformed to final structure of levofloxacin through further 4 steps (i.e., incorporation of N-methylpiperazine unit through the nucleophilic aromatic substitution reaction, activation, displacement of primary alcohol, and ester hydrolysis) [45].

The ofloxacin, (-)-9-3-methyl-10-(4-methyl-1-piperazinyl)-7H-pyrido(1,2,3-de)-1,4-benzoxazin-7-oxo-6-carboxylic acid, has a slightly more complex structure as compared to drugs, the key dissimilarity was the presence of Me-substituted oxyethylene bridge between the C8-position of the quinolone system and nitrogen atom as well as the presence of a Me substituent at C4-position of the piperazine ring. To synthesize the oxo-analog of Gerster intermediate, Hayakawa et al. [46] started the synthesis with selective displacement of fluorine *ortho* to the NO$_2$ group in 2,3,4-trifluoro-1-nitrobenzene with KOH in dimethylsulfoxide to provide the phenol in 29% yield (Scheme 2.25). Then, the phenol was alkylated with α-CH$_3$COCH$_2$Cl using KI and K$_2$CO$_3$ to provide the ether of hydroxyacetone in moderate yield. The reduction of NO$_2$ group with Ra-Ni and H$_2$ followed by concomitant cyclization of formed aniline methyl ketone provided difluorobenzoxazine in 90% yield. The benzoxazine was then transformed to quinolone by Gould–Jacobs approach. First, the malonate, synthesized by condensation of benzoxazine with diethyl ethoxymethylenemalonate at 145°C, underwent a cyclization using PPE (polyphosphoric ester) as a dehydrating agent at 140 to 145°C to provide the tricyclic core in 64% yield over two steps. The ethyl ester of quinolone was hydrolyzed in a mixture of HCl and AcOH to carboxylic acid in 94% yield. The selective displacement of C-8 fluorine with N-methylpiperazine in dimethylsulfoxide at 100°C provided (+)-ofloxacin in 62% yield. However, unlike any of the cases where the reaction was executed at high temperatures, this reaction was accomplished utilizing polyphosphoric acid [47–51].

2.7 Reduction of NO$_2$ group for the synthesis of seven-membered heterocycles

The known indole, easily accessible on multigram scale [52] from Fischer indole synthesis involving phenylhydrazine and *o*-nitroacetophenone, was reacted with (COCl)$_2$ [53] in ether to afford the Friedel–Crafts acid

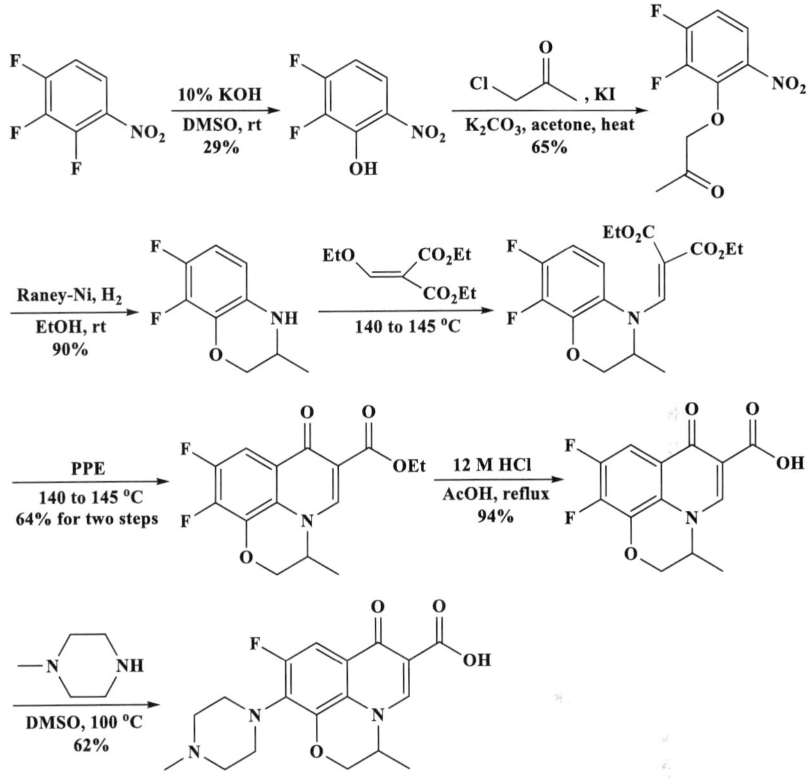

Scheme 2.25

halide product, which underwent an alcoholysis in EtOH. The product was precipitated from solution in pure form during both operations. The protection of ketone functionality as dithiolane by acid-catalyzed reaction with 1,2-ethanedithiol provided NO_2 compound in 78% overall yield. The reaction of NO_2 compound with Raney-Ni followed by heating in dioxane and AcOH allowed (without purification) the reduction of NO_2 moiety, hydrogenolysis of dithiolane to a methylene group [54] and cyclization by an attack of in situ generated aniline on the ester to give the paullone in 54% yield (Scheme 2.26) [55]. Study of this reaction as it progressed showed that the cyclization was the final step. The pathway was concise and efficient which used an easily accessible starting material, took the benefit of various reaction steps in one-pot and needed no chromatography [56–59]. Additionally, since the Fischer indole synthesis was moved to the beginning of the path, this approach offers a greater degree of certainty regarding

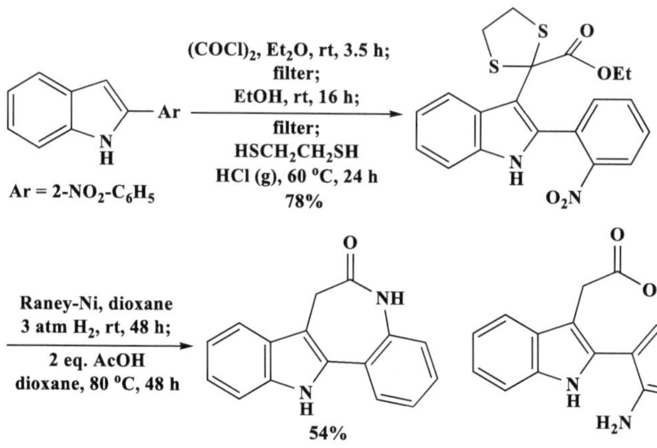

Scheme 2.26

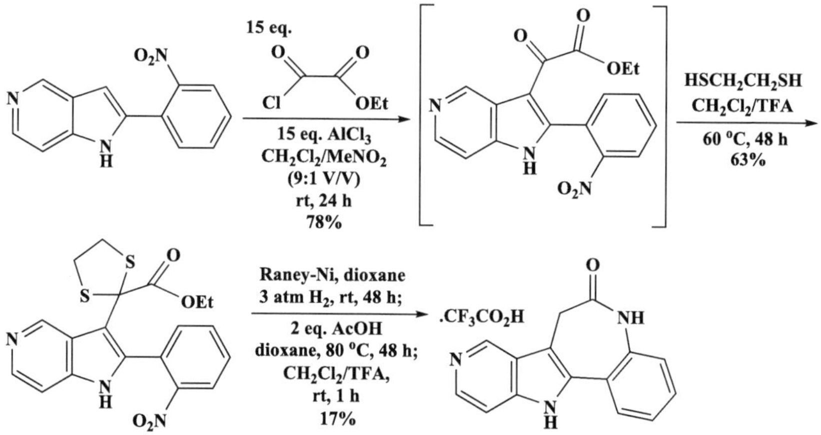

Scheme 2.27

versatility. One can prepare the indole with the required substitution pattern on both aromatic rings at the starting of the sequence and then synthesized ε-lactam (which was important for activity) through a novel, short, two-pot sequence, in place of necessitating that potential modifications to the paullone core be compatible with the often harsh conditions associated with Fischer indole synthesis in the final step.

The azaindole was converted to 9-azapaullone (Scheme 2.27). The azaindole was transformed to NO_2 compound (in addition to the diethylketal if EtOH was utilized to stop the reaction) with the aid of aluminum chloride using excess of acid chloride. The protection of ketone functionality as its dithiolane derivative (in overall good yield from azaindole) allowed the similar reduction, hydrogenolysis, and cyclization sequence which was utilized to synthesize the desired azapaullone, which was isolated as its trifluoroacetic acid salt because of the poor solubility of free base in organic solvents [60].

In this method, the unprotected aldehyde was cyclized to provide the pyrrolobenzodiazepine system by the reduction of NO_2 group. This approach has resemblances to the cyclization of N-(2-nitrobenzoyl)pyrrolidine carboxylates. The reductive cyclization was primarily achieved using Pd-catalyzed hydrogenation on charcoal [61] of NO_2 functionality on the aromatic ring. Thurston and Langley [62] utilized this methodology and found difficulties of over reduction of the N10-C11 imine bond to yield the biologically inactive secondary amine instead. Rojas-Rousseau and Langlois [63] utilized reductive cyclization method in the formation of (+)-porothramycin and anthramycin. The pyrrolobenzodiazepine system was prepared by selectively reducing the NO_2 aldehyde with Raney nickel catalyst. Rojas–Rousseau and Langlois [63] have described the formation of an analogue of porothramycin (Scheme 2.28), which used Raney nickel NO_2 reduction. The first step was the synthesis of (5S)-5-(ethoxy-ethoxymethyl)pyrrolidin-2-one according to literature approaches [64]. After deprotonation, this compound was reacted with 2-nitrobenzoyl chloride to afford the imide in 95% yield. The unstable α-hydroxy-2-nitrobenzamides were prepared by reduction with diisobutylaluminium hydride at -78°C in toluene, and were quantitatively transformed to more stable α-methoxy-2-nitrobenzamides. The primary alcohol was protected again with an acetyl group and the OCH_3 group was removed to synthesize the enamide in high yield by heating in toluene using QCS (quinolinium camphorsulfonate) as a catalyst. The enamide was transformed into a highly versatile unsaturated aldehyde functionality quantitatively through a Vilsmeier-Haack (V-H) reaction. The aldehyde was reacted with tetramethyl-methylenediphosphonate using 1 eq. *n*-butyl lithium at 0°C to deliver the acetate as a single diastereoisomer in 87% yield. The saponification of acetate with $Ba(OH)_2$ provided primary alcohol, which on Swern oxidation with N,N-di-*i*-propylethylamine as a base afforded aldehyde in almost quantitative yield (99%). The porothramycin analogue was constructed by reductive cyclization using Ra-nickel followed by reaction with MeOH and small amounts of trifluoroacetic acid.

A methodology was described by Langois et al. [65] towards a different analogue in which the A ring possess a methylenedioxy

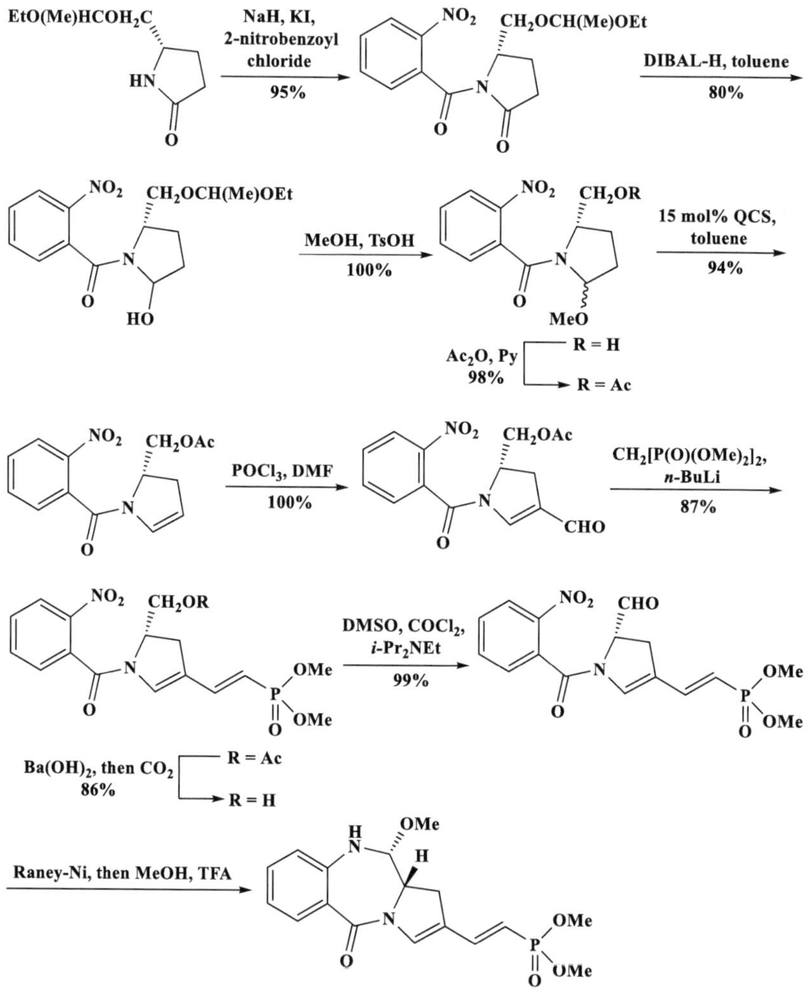

Scheme 2.28

group (Scheme 2.29). This was reacted with (5S)-5-(ethoxy-ethoxymethyl)pyrrolidin-2-one and then NO_2 acetylated aldehyde. The acetyl derivative was prepared by the condensation of aldehyde with lithiated anion of diethyl[2-(dimethylamino)-2-oxoethyl]phosphonate. The acyl protection of primary alcohol was removed by alkaline hydrolysis to provide the hydrogen derivative, followed by an oxidation under Swern conditions utilizing Hünig's base to afford the pyrrolobenzodiazepine precursor without any racemization. The reduction of aromatic -NO_2

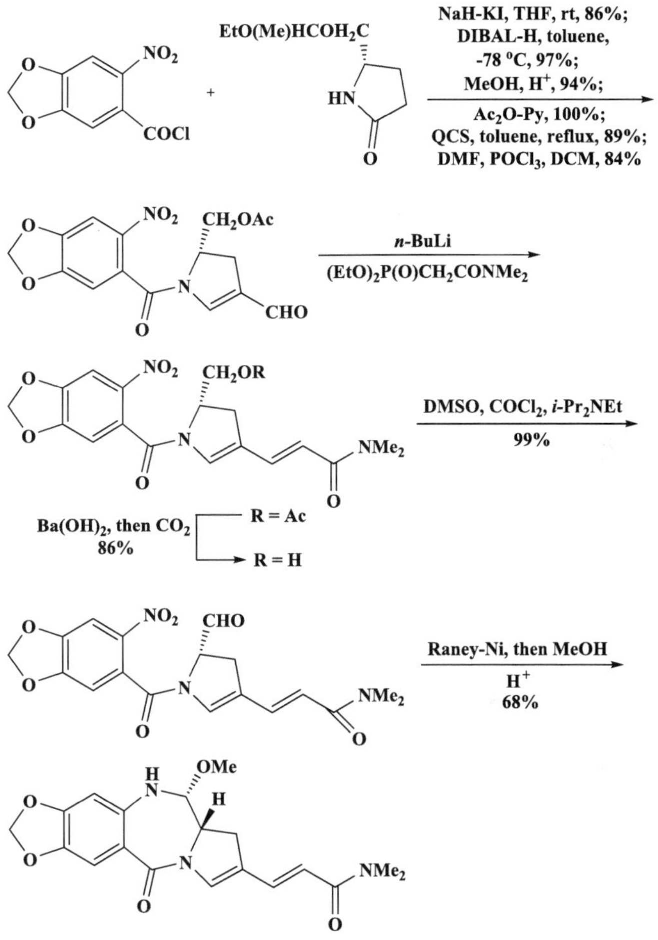

Scheme 2.29

functionality with Ra-nickel provided pyrrolobenzodiazepine imine, which was not purified at this stage but was transformed to crystalline carbinolamine methyl ether by weak acid–methanol treatment.

Long-term clozapine therapy exhibited a greater efficiency in the treatment of schizophrenia compared to typical antipsychotic medicines [66]. It is serotonin antagonist that shows a strong binding to 5-HT2A/2C receptor, subtype and a partial agonist for the 5-HT1A receptor. The clozapine is also a strong antagonist at diverse subtypes of histaminergic, adrenergic, and cholinergic receptors. In spite of its efficiency and lack

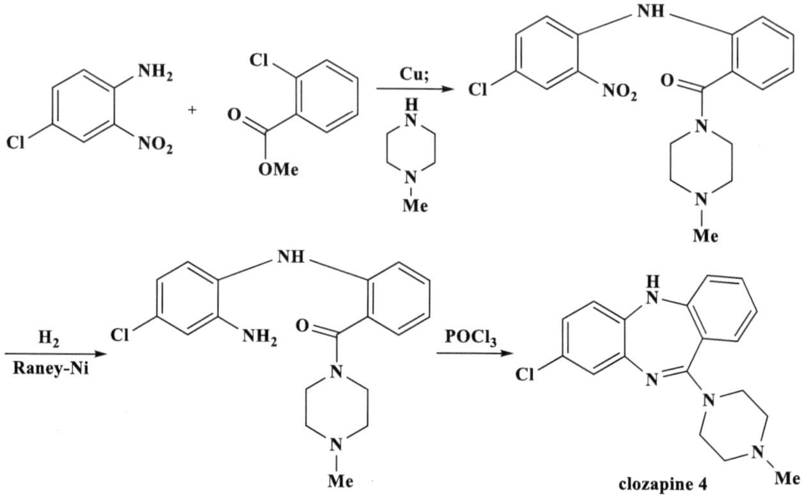

Scheme 2.30

of extra pyramidal side effects, its use has been limited due to its tendency to cause agranulocytosis [67] and other toxic effects [68] including cardiac and liver poisonousness [69,70]. The clozapine, 8-chloro-11-(4-methyl-1-piperazinyl)-5H-dibenzo[b,e][1,4]diazepine, was prepared by two approaches. First, the diphenyl amine with o-chlorobenzoic acid methyl ester was synthesized from 4-chloro-2-nitroaniline in the presence of Cu filings. The diphenyl amine was then converted to amide with N-methylpiperazine. The reduction of this NO_2 compound with Raney-Ni and subsequent reaction of amino product with phosphorus oxytrichloride synthesized clozapine by heterocyclization (Scheme 2.30) [71–73].

The 11-(piperazin-1-yl)-5H-dibenzo[b,e][1,4]diazepine was prepared on kilo scale without any chromatographic purification. The key steps Ullmann condensation, catalytic hydrogenation, and catalyzed cyclization were involved. The key intermediate N-(2-nitrophenyl) anthranilic acid was synthesized by Ullmann condensation reaction between 2-fluoronitrobenzene and anthranilic acid. The NO_2 group in anthranilic acid was reduced by catalytic hydrogenation applying W-2 Ra-Ni to provide the N-(2-aminophenyl) anthranilic acid. The cyclization of N-(2-aminophenyl) anthranilic acid was performed in dimethylformamide utilizing sulfuric acid in catalytic amounts. The 5H-dibenzo[b,e][1,4]diazepine-11(10H)-one was condensed with piperazine applying titanium tetrachloride to obtain the 11-piperazinyl-5H-dibenzo[b,e][1,4]diazepine (Scheme 2.31) [74].

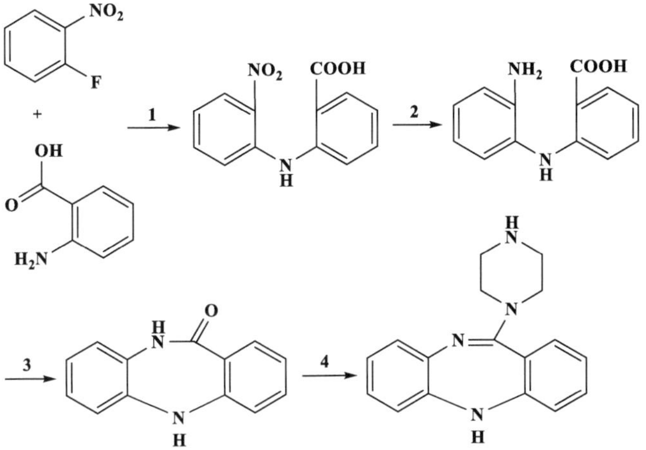

Reagents and conditions: (1) DMF, K$_2$CO$_3$, Cu powder, reflux, 12 h, 85%, (2) H$_2$, Raney-Ni, 70 psi, 2 h, 83%, (3) DMF, H$_2$SO$_4$, reflux, 7-8 h, 81%, (4) anisole, piperazine, TiCl$_4$, reflux, 7-8 h, 63%.

Scheme 2.31

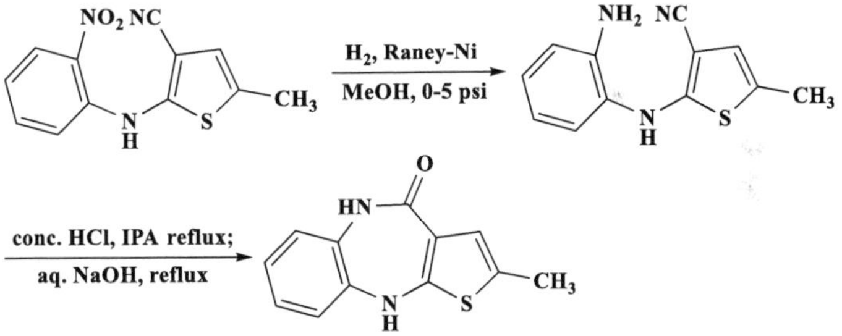

Scheme 2.32

The cyclic amide was prepared by: (i) reducing the nitro thiophene carbonitrile with Raney-Ni at 0-5 psi pressure to give the amino thiophene carbonitrile; and (ii) transforming the amino thiophene carbonitrile to cyclic amide. The synthetic method followed for the synthesis of cyclic amide is shown in Scheme 2.32 [74].

It was reported that the reduction with Zn and EtOAc led to complete over reduction of the imine bond [75]. It is frequently realized that when reductive conditions are applied with NO$_2$ derivatives, the amine intermediate (Scheme 2.33) and the carbonyl underwent an immediate spontaneous intramolecular cyclization. Langlois and Andriamialisoa [76] reported that the amino derivative was isolated when Raney-Ni was

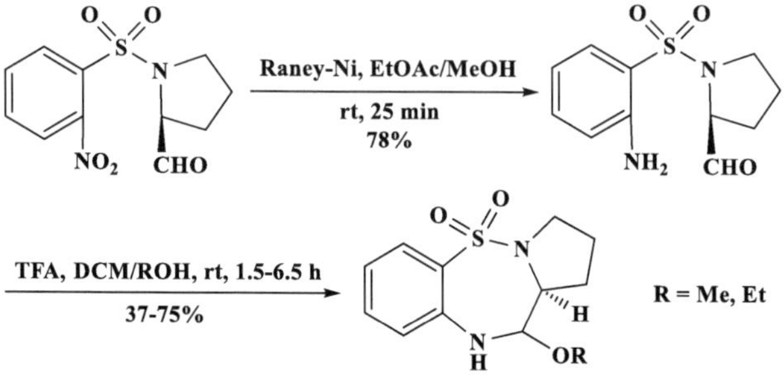

Scheme 2.33

utilized as a reducing agent with NO$_2$ precursor and subsequent cyclization afforded desired compound i.e. the sulfur analogue of natural antitumor antibiotic PBD abbeymycin.

References

[1] (a) Kaur N. Palladium catalysts: synthesis of five-membered *N*-heterocycles fused with other heterocycles. Catal Rev 2015;57:1–78; (b) Kaur N. Synthesis of seven and higher-membered heterocycles using ruthenium catalysts. Synth Commun 2019;49:617–61; (c) Kaur N. Applications of palladium dibenzylideneacetone as catalyst in the synthesis of five-membered *N*-heterocycles. Synth Commun 2019;49:1205–30; (d) Kaur N. Copper catalyzed synthesis of seven and higher-membered heterocycles. Synth Commun 2019;49:879–916; (e) Kaur N. Ionic liquid assisted synthesis of *S*-heterocycles. Phosphorus, Sulfur, Silicon Relat Elem 2019;194:165–85; (f) Kaur N. Nickel catalysis: six membered heterocycle syntheses. Synth Commun 2019;49:1103–33; (g) Kaur N, Tyagi R, Kishore D. Expedient protocols for the installation of 1,5-benzoazepino based privileged templates on 2-position of 1,4-benzodiazepine through a phenoxyl spacer. J Heterocycl Chem 2016;53:643–6.

[2] (a) Kaur N. Role of microwaves in the synthesis of fused five-membered heterocycles with three *N*-heteroatoms. Synth Commun 2015;45:403–31; (b) Kaur N, Kishore D. Microwave-assisted synthesis of seven- and higher-membered *O*-heterocycles. Synth Commun 2014;44:2739–55; (c) Kaur N, Kishore D. Microwave-assisted synthesis of six-membered *S*-heterocycles. Synth Commun 2014;44:2615–44; (d) Kaur N. Ultrasound-assisted green synthesis of five-membered *O*- and *S*-heterocycles. Synth Commun 2018;48:1715–38; (e) Kaur N, Tyagi R, Srivastava M, Kishore D. Application of dimethylaminomethylene ketone in heterocycles synthesis: synthesis of 2-(isoxazolo, pyrazolo and pyrimido) substituted analogs of 1,4-benzodiazepin-5-carboxamides linked through an oxyphenyl bridge. J Heterocycl Chem 2014;51:E50–4; (f) Kaur N. A new approach to anti-HIV chemotherapy devised by linking the vital fragments of active RT inhibitors such as etravirine to the molecular framework of anti-HIV prone privileged nucleus of 1,4-benzodiazepine as possible substitute to 'HAART'. Int J Pharm Biol Sci 2013;4:309–17.

[3] (a) Kaur N. Mercury-catalyzed synthesis of heterocycles. Synth Commun 2018;48:2715–49; (b) Kaur N. Photochemical irradiation: seven- and higher-membered *O*-heterocycles. Synth Commun 2018;48:2935–64; (c) Kaur N. Seven-membered *N*-heterocycles: metal and non-metal assisted synthesis. Synth Commun 2019;49:987–1030;

(d) Kaur N, Bhardwaj P, Devi M, Verma Y, Grewal P. Synthesis of five-membered O,N-heterocycles using metal and non-metal. Synth Commun 2019;49:1345–84; (e) Kaur N. Expedient protocol for the installation of thiadiazole on 2-position of 1,4-benzodiazepin-5-carboxamide through a phenoxyl spacer. Int J Pharm Biol Sci 2013;4:366–73; (f) Anand A, Kaur N, Kishore D. An efficient one pot protocol to the annulation of face 'd' of benzazepinone ring with pyrazole, isoxazole and pyrimidine nucleus through the corresponding oxoketene dithioacetal derivative. Adv Chem 2014:1–5.

[4] (a) Kaur N. Photochemical-mediated reactions in five-membered O-heterocycles synthesis. Synth Commun 2018;48:2119–49; (b) Kaur N. Synthetic routes to seven and higher membered S-heterocycles by use of metal and nonmetal catalyzed reactions. Phosphorus, Sulfur, Silicon Relat Elem 2019;194:186–209; (c) Kaur N. Synthesis of six-membered N-heterocycles using ruthenium catalysts. Catal Lett 2019;14:1513–39; (d) Kaur N. Green synthesis of three to five-membered O-heterocycles using ionic liquids. Synth Commun 2018;48:1588–613; (e) Kaur N. Synthesis of seven- and higher-membered nitrogen-containing heterocycles using photochemical irradiation. Synth Commun 2018;48:2815–49; (f) Kaur N. Ruthenium-catalyzed synthesis of five-membered O-heterocycles. Inorg Chem Commun 2018;99:82–107; (g) Kaur N, Sharma P, Sirohi R, Kishore D. Microwave assisted synthesis of 2-hetryl amino substituted novel analogues of 1,4-benzodiazepine-5-piperidinyl carboxamides. Arch Appl Sci Res 2012;4:2256–60.

[5] (a) Kaur N. Solid-phase synthesis of sulfur-containing heterocycles. J Sulfur Chem 2018;39:544–77; (b) Kaur N. Photochemical reactions as key steps in five-membered N-heterocycles synthesis. Synth Commun 2018;48:1259–84; (c) Kaur N. Recent developments in the synthesis of nitrogen-containing five-membered polyheterocycles using rhodium catalysts. Synth Commun 2018;48:2457–74; (d) Kaur N. Synthesis of five-membered heterocycles containing nitrogen heteroatom under ultrasonic irradiation. Mini Rev Org Chem 2019;16:481–503; (e) Kaur N. Cobalt-catalyzed C-N, C-O, C-S bond formation: synthesis of heterocycles. J Iran Chem Soc 2019;16:2525–53; (f) Rao K, Tyagi R, Kaur N, Kishore D. An expedient protocol to the synthesis of benzo(b)furans by palladium induced heterocyclization of corresponding 2-allylphenols containing electron rich and electron capturing substituents in the arene ring. J Chem 2013:1–5.

[6] (a) Kaur N. Ionic liquids: a versatile medium for the synthesis of six-membered two nitrogen containing heterocycles. Curr Org Chem 2019;23:76–96; (b) Kaur N. Photochemical reactions for the synthesis of six-membered O-heterocycles. Curr Org Synth 2018;15:298–320; (c) Kaur N. Perspectives of ionic liquids applications for the synthesis of five and six-membered O,N-heterocycles. Synth Commun 2018;48:473–95; (d) Kaur N. Metal and non-metal catalysts in the synthesis of five-membered S-heterocycles. Curr Org Synth 2019;16:258–75; (e) Kaur N, Bhardwaj P, Devi M, Verma Y, Ahlawat N, Grewal P. Ionic liquids in the synthesis of five-membered N,N-, N,N,N- and N,N,N,N-heterocycles. Curr Org Chem 2019;23:1214–38; (f) Kaur N, Kishore D. Synthesis of oxadiazolo, pyrimido, imidazolo and benzimidazolo containing derivatives of 1,4-benzodiazepin-5-(4'-methylpiperazinyl)-carboxamide through phenylamino spacer. Synth Commun 2014;44:2789–96; (g) Kaur N, Aditi, Kishore D. A facile synthesis of face 'D' quinolino annulated benzazepinone analogues with its quinoline framework appended to oxadiazole, triazole and pyrazole heterocycles. J Heterocycl Chem 2016;53:457–60.

[7] Sinclair A, Stockman RA. Thirty-five years of synthetic studies directed towards the histrionicotoxin family of alkaloids. Nat Prod Rep 2007;24:298–326.

[8] Sridharan V, Suryavanshi PA, Menendez JC. Advances in the chemistry of tetrahydroquinolines. Chem Rev 2011;111:7157–259.

[9] Sun B-F. Total synthesis of natural and pharmaceutical products powered by organocatalytic reactions. Tetrahedron Lett 2015;56:2133–40.

[10] Mittapelli V, Reddy NR, Reddy PP. Synthesis of novel fused pyrazolo[4,3-d][1,2,3]triazin-4-ones: application of diazonium ion induced heterocyclisation in the synthesis of novel heterocycles. Int J Biol Pharm Res 2013;4:592–7.
[11] Heaney F. The nitro group in organic synthesis, New York: John Wiley - VCH; 2001. N. Ono. (Ed.)ISBN 0-471-31611-3, 2528.
[12] Sabatier P, Senderens JB. The nitro group in organic synthesis. N Ono (Ed). Compt Rend 1897;124:1359.
[13] Barrett AGM, Kohrt JT. Model studies on the synthesis of hennoxazole A. Synlett 1995;5:415–16.
[14] Stowell GJ. Original synthesis of 7-amino-5,8-dimethyl-6-(thiophenyl)isoquinoline. PhD Dissertation. Davis: University of California; 1981. p. 84–108.
[15] Miller RB, Stowell JG, Jenks CW, Farmer SC, Wujcik CE, Olmstead MM. A new ring system: 3H-pyrazolo[3,4-h]isoquinoline. An unexpected product from diazotization of an aminoisoquinoline. Chem Commun 1996;24:2711–12.
[16] March J. Nitrite as leaving group. Advanced organic chemistry: reactions, mechanism, and structure. 3rd Ed. New York: Wiley; 1985. p. 587.
[17] Johnson TA, Curtis MD, Beak P. Highly diastereoselective and enantioselective carbon-carbon bond formations in conjugate additions of lithiated N-Boc allylamines to nitroalkenes: enantioselective synthesis of 3,4- and 3,4,5-substituted piperidines including (-)-paroxetine. J Am Chem Soc 2001;123:1004–5.
[18] Johnson TA, Jang DO, Slafer BW, Curtis MD, Beak P. Asymmetric carbon-carbon bond formations in conjugate additions of lithiated N-Boc allylic and benzylic amines to nitroalkenes: enantioselective synthesis of substituted piperidines, pyrrolidines, and pyrimidinones. J Am Chem Soc 2002;124:11689–98.
[19] Fitch RW, Luzzio FA. Ultrasound in natural products synthesis: applications to the synthesis of histrionicotoxin via nitroalkanal acetals. Ultrason Sonochem 1997;4:99–107.
[20] Luzzio FA, Fitch RW. Formal synthesis of (+)- and (-)-perhydrohistrionicotoxin: a "double Henry"/enzymatic desymmetrization route to the Kishi lactam. J Org Chem 1999;64:5485–93.
[21] Gomez MFE. Modeling nonlinearities in agricultural factor markets. PhD Thesis, North Carolina State University, Onel, Gulcan. 2013;95:527-8.
[22] Chabert JFD, Chatelain G, Pellet-Rostaing S, Bouchu D, Lemaire M. Benzo[b]thiophene as a template for substituted quinolines and tetrahydroquinolines. Tetrahedron Lett 2006;47:1015–18.
[23] Tsang AS-K, Ingram K, Keiser J, Hibbert DB, Todd MH. Enhancing the usefulness of cross dehydrogenative coupling reactions with a removable protecting group. Org Biomol Chem 2013;11:4921–4.
[24] Banwell MG, Jones MT, Reekie TA, Schwartz BD, Tan SH, White LV. Raney® cobalt - an under utilised reagent for the selective cleavage of C-X and N-O bonds. Org Biomol Chem 2014;12:7433–44.
[25] Ng PY, Masse CE, Shaw JT. Cycloaddition reactions of imines with 3-thiosuccinic anhydrides: synthesis of the tricyclic core of martinellic acid. Org Lett 2006;8:3999–4002.
[26] Nyerges M, Fejes I, Toke L. An intermolecular 1,3-dipolar cycloaddition approach to the tricyclic core of martinelline and martinellic acid. Tetrahedron Lett 2000;41:7951–4.
[27] Nyerges M, Fejes I, Toke LA. Convenient synthesis of the pyrrolo[3,2-c]quinoline core of martinelline alkaloids. Synthesis 2002;13:1823–8.
[28] Stetter H, Schwarz M. Synthese des 4.5.9.10-tetraaza-pyrens. Chem Ber 1957;90:1349–51.
[29] Sienkowska MJ, Farrar JM, Zhang F, Kusuma S, Heiney PA, Kaszynski P. Liquid crystalline behavior of tetraaryl derivatives of benzo[c]cinnoline, tetraazapyrene, phenanthrene, and pyrene: the effect of heteroatom and substitution pattern on phase stability. J Mater Chem 2007;17:1399–411.

[30] Bella S, Brown D and Terrett NK. 1991. EP Patent 463756.
[31] Yan-Fang Z, Xin Z, Jiao-Yue C, Shu-Chun G, Ping G. Syntheses and vasodilatory activities of new pyrazolo[4,3-d]pyrimidin-7-ones. Chem Res Chin 2006; 22:468–73.
[32] Buyck T, Wang Q, Zhu J. Catalytic enantioselective Michael addition of α-aryl-α-isocyanoacetates to vinyl selenone: synthesis of α,α-disubstituted α-amino acids and (+)- and (-)-trigonoliimine A. Angew Chem, Int Ed 2013;52:12714–18.
[33] Zhu S, Yu S, Wang Y, Ma D. Organocatalytic Michael addition of aldehydes to protected 2-amino-1-nitroethenes: the practical synthesis of oseltamivir (tamiflu) and substituted 3-aminopyrrolidines. Angew Chem, Int Ed 2010;49:4656–60.
[34] Gualtierotti JB, Pasche D, Wang Q, Zhu J. Phosphoric acid catalyzed desymmetrization of bicyclic bislactones bearing an all-carbon stereogenic center: total synthesis of (-)-rhazinilam and (-)-leucomidine B. Angew Chem, Int Ed 2014;53:9926–30.
[35] Cannizzo S, Guerrera F, Siracusa MA. Synthesis of substituted [1]benzothieno[2,3-b]pyrazines. J Heterocycl Chem 1990;27:2175–9.
[36] Smits RA, Lim HD, Hanzer A, Zuiderveld OP, Guaita E, Adami M, Coruzzi G, Leurs R, de Esch IJP. Fragment based design of new H4 receptor-ligands with anti-inflammatory properties in vivo. J Med Chem 2008;51:2457–67.
[37] Clarke P, Moorehouse A. The synthesis of some 6-substituted 1,2,3,4-tetrahydroquinoxalines. J Chem Soc 1963;0:4763–7.
[38] Turski L, Schneider HH, Neuhaus R, McDonald F, Jones GH, Lofberg B, Schweinfurth H, Huth A, Kruger M, Ottow E. Phosphonate quinoxalinedione AMPA antagonists. Restor Neurol Neurosci 2000;17:45–59.
[39] Temple C, Rose JD, Elliot RD, Montgomery JA. Synthesis of potential antimalarial agents. II. 6,8-Disubstituted pyrido[2,3-b]pyrazines. J Med Chem 1968;11:1216–18.
[40] Reynolds RC, Srivastava S, Ross LJ, Suling WJ, White EL. A new 2-carbamoyl pteridine that inhibits mycobacterial FtsZ. Bioorg Med Chem Lett 2004;14:3161–4.
[41] Elliot RD, Temple C, Montgomery JA. Potential folic acid antagonists. II. Deaza analogs of methotrexate. II. 2,4-Diamino-6-methyl-3-deazapteridine. J Org Chem 1966;31:1890–4.
[42] Mathew B, Srivastava S, Ross LJ, Suling WJ, White EL, Woolhiser LK, Lenaerts AJ, Reynolds RC. Novel pyridopyrazine and pyrimidothiazine derivatives as FtsZ inhibitors. Bioorg Med Chem 2011;19:7120–8.
[43] Atarashi S, Tsurumi H, Fujiwara T, Hayakawa I. Asymmetric reduction of 7,8-difluoro-3-methyl-2H-1,4-benzoxazine. Synthesis of a key intermediate of (S)-(-)-ofloxacin (DR-3355). J Heterocycl Chem 1991;28:329–31.
[44] Hayakawa IDS, Tanaka YDS. 1986. Tricyclic compounds, a process for their production and pharmaceutical compositions containing said compounds. Eur Patent EP 101 829 B1.
[45] Baumann M, Baxendale IR. An overview of the synthetic routes to the best selling drugs containing 6-membered heterocycles. Beilstein J Org Chem 2013;9:2265–319.
[46] Hayakawa I, Tanaka Y and Hiramitsu T. 1982. Preparation of levofloxacin hemihydrate. Eur Patent Appl 47005.
[47] Matsumoto J, Takase Y and Nishimura Y. 1983. Benzoxazine derivatives. US Patenet 4.382.892.
[48] Egawa H, Miyamoto T, Matsumoto J. A new synthesis of 7H-pyrido(1,2,3-d,e)(1,4)benzoxazine derivatives including an antibacterial agent, ofloxacin. Chem Pharm Bull 1986;34:4098–102.
[49] Atarashi S, Yokohama S, Yamazaki K, Sekano K, Imamura M, Hayakawa I. Synthesis and antibacterial activities of optically active ofloxacin and its fluoromethyl derivative. Chem Pharm Bull 1987;35:1896–908.
[50] Andriole VT. The quinolones: past, present, and future. Clin Infect Dis 2005; 41:113–19.

[51] Appelbaum PC, Gillespie SH, Burley CJ, Tillotson GS. Antimicrobial selection for community-acquired lower respiratory tract infections in the 21st century: a review of gemifloxacin. Int J Antimicrob Agents 2004;23:533–46.
[52] MacPhillamy HB, Dziemian RL, Lucas RA, Kuehne ME. The alkaloids of *Tabernanthe iboga*. Part VI.1 The synthesis of the selenium dehydrogenation products from ibogamine. J Am Chem Soc 1958;80:2172–8.
[53] Han Q, Dominguez C, Stouten PFW, Park JM, Duffy DE, Galemmo RA, Rossi KA, Alexander RS, Smallwood AM, Wong PC, Wright MM, Luetten JM, Knabb RM, Wexler RR. Design, synthesis, and biological evaluation of potent and selective amidino bicyclic factor Xa inhibitors. J Med Chem 2000;43:4398–415.
[54] Melhado LL, Brodsky JL. Synthesis of 4-azidoindole-3-acetic acid, a photoprobe causing sustained auxin activity. J Org Chem 1988;53:3852–5.
[55] Opatz T, Ferenc D. Synthesis of the CDK-inhibitor paullone by cyclization of a deprotonated α-aminonitrile. Synthesis 2008;24:3941–4.
[56] Baudoin O, Cesario M, Guenard D, Gueritte F. Application of the palladium-catalyzed borylation/Suzuki coupling (BSC) reaction to the synthesis of biologically active biaryl lactams. J Org Chem 2002;67:1199–207.
[57] Joucla L, Popowycz F, Lozach O, Meijer L, Joseph B. Access to paullone analogues by intramolecular Heck reaction. Helv Chim Acta 2007;90:753–63.
[58] Bremner JB, Sengpracha W. An iodoacetamide-based free radical cyclisation approach to the 7,12-dihydro-indolo[3,2-*d*][1]benzazepin-6(5*H*)-one (paullone) system. Tetrahedron 2005;61:5489–98.
[59] Henry N, Blu J, Beneteau V, Merour J-Y. New route to the 5,12-dihydro-7*H*-benzo[2,3]azepino[4,5-*b*]indol-6-one core via a tin-mediated indole synthesis. Synthesis 2006;22:3895–901.
[60] Power DP, Lozach O, Meijer L, Grayson DH, Connon SJ. Concise synthesis and CDK/GSK inhibitory activity of the missing 9-azapaullones. Bioorg Med Chem Lett 2010;20:4940–4.
[61] Artico M, de Martino G, Giuliano R, Massa S, Porretta GC. Synthesis of 5*H*-pyrrolo[2,1-*c*][1,4]benzodiazepine and some of its derivatives related to anthramycin. J Chem Soc, Chem Commun 1969;12:671-671.
[62] Thurston DE, Langley DR. Synthesis and stereochemistry of carbinolamine-containing pyrrolo[1,4]benzodiazepines by reductive cyclization of *N*-(2-nitrobenzoyl)pyrrolidine-2-carboxaldehydes. J Org Chem 1986;51:705–12.
[63] Rojas-Rousseau A, Langlois N. Synthesis of a new structural analogue of (+)-porothramycin. Tetrahedron 2001;57:3389–95.
[64] Saijo S, Wada M, Himizu J, Ishida A. Heterocyclic prostaglandins. V. Synthesis of (12*R*,15*S*)-(-)-11-deoxy-8-azaprostaglandine and related compounds. Chem Pharm Bull 1980;28:1449–58.
[65] Langlois N, Rojas-Rousseau A, Gaspard C, Werner GH, Darro F, Kiss R. Synthesis and cytotoxicity on sensitive and doxorubicin-resistant cell lines of new pyrrolo[2,1-*c*][1,4]benzodiazepines related to anthramycin. J Med Chem 2001;44:3754–7.
[66] Ravanic DB, Dejanovic SM, Janjic V. Effectiveness of clozapine, haloperidol and chlorpromazine in schizophrenia during a five-year period. Arq Neuropsiquiatr 2009;67:195–202.
[67] Atkin K, Kendall F, Gould D, Freeman H, Liberman J, Osullivan D. Neutropenia and agranulocytosis in patients receiving clozapine in the UK and Ireland. Br J Psychiatry 1996;169:483–8.
[68] Flanagan RJ, Dunk L. Haematological toxicity of drugs used in psychiatry. Hum Psychopharmacol 2008;23:27–41.
[69] Layland JJ, Liew D, Prior DL. Clozapine-induced cardiotoxicity: a clinical update. Med J Aust 2009;190:190–2.

[70] Markowitz JS, Grinberg R, Jackson C. Marked liver enzyme elevations with clozapine. J Clin Psychopharmacol 1997;17:70–1.
[71] Wander SAA. 1963. A process for preparing amidines from the series of 5-dibenzo[b,e][1,4]diazepine. Fr. Patent 1,334,944.
[72] Hunziker F and Shmutz J. 1970. Basic substituted dibenzodiazepines and dibenzothiazepines. US Patent 3,539,573.
[73] Wander SAA. 1961. Br Patent 980,853.
[74] Kalhapure RS, Patil BP, Jadhav MN, Kawle LA, Wagh SB. Synthesis of 11-(piperazin-1-yl)-5H-dibenzo[b,e][1,4]diazepine on kilo scale. E J Chem 2011;8:1747–9.
[75] Cherney RJ, Duan JJW, Voss ME, Chen L, Wang L, Meyer DT, Wasserman ZR, Hardman KD, Liu R-Q, Covington MB. Design, synthesis, and evaluation of benzothiadiazepine hydroxamates as selective tumor necrosis factor-α converting enzyme inhibitors. J Med Chem 2003;46:1811–23.
[76] Langlois N, Andriamialisoa RZ. Synthesis of sulfonamide analogs of the pyrrolo[1,4]benzodiazepine antibiotic abbeymycin. Heterocycles 1989;29:1529–36.

CHAPTER 3

Synthesis of heterocycles from cyanide, oxime, and azo compounds using Raney nickel

3.1 Introduction

Heterocycles mimic the major pharmaceutical products and natural products with biological activities. Various significant advancements have been made against diseases by developing and examining new structures, which are generally heteroaromatic derivatives. Additionally, there are a number of heterocyclic natural products such as alkaloids, antibiotics, cardiac glycosides, and pesticides having a lot of importance for human and animal health. Therefore, natural models have been followed for designing and constructing weed killers, pharmaceuticals, pesticides, rodenticides, and insecticides. An important part of such biologically active compounds is formed by heterocycles [1–4]. These compounds play an important role in biochemical processes and they also worked as side groups of most typical and important components of living cells. Other important practical uses of heterocycles are that they act as modifiers and additives in a wide range of industries such as reprography, plastics, cosmetics, vulcanization accelerators, solvents, antioxidants, and information storage. Finally, heterocyclic chemistry is an inexhaustible source of novel compounds as an applied science. The compounds with most diverse chemical, physical, and biological properties are provided by designing a wide range of combinations of carbon, hydrogen, and heteroatoms [5–7].

The *N*-containing heterocycles (azaheterocycles) are present in drugs, natural products, and other biologically active molecules. Therefore, considerable synthetic methods have been established for the construction of azaheterocycles [8–12].

Nickel was the first metal catalyst to be used to perform the *N*-alkylation of amines with other amines being the source of the electrophile. This

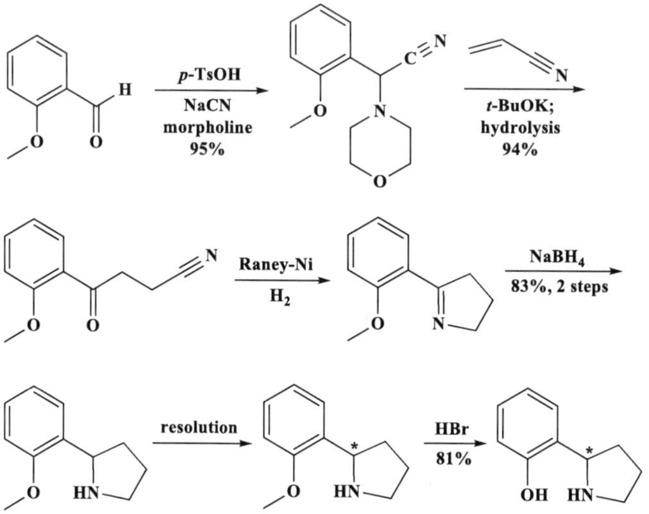

Scheme 3.1

catalytic ability was found when Raney nickel was used as a catalyst to perform the reduction of nitrile derivatives [13].

3.2 Synthesis of heterocycles from cyanide compounds

3.2.1 Synthesis of five-membered heterocycles from cyanide compounds

The γ-ketonitrile was prepared by transformation of O-methoxybenzaldehyde in a two-step Stetter reaction [14]. This less well-known method of umpolung is very mild and convenient and should get more attention in organic synthesis. At slightly increased temperatures, the deprotection of ketone proceeded through a mild acid catalysis. The nitrile was reduced to an amine by a Raney nickel-catalyzed hydrogenation. The amine was not isolated and synthesized the cyclized imine in situ, which on reduction with sodium borohydride afforded cyclic amine (79% yield, 2 steps). Dutch resolution methods were used for the resolution of amine into its enantiomers using a family of mandelic acids [15,16]. The phenol was formed in 81% isolated yield when the methoxy moiety was demethylated by refluxing in conc. hydrogen bromide. The chiral HPLC determined the ee of synthons and was found to be higher than 97% (Scheme 3.1).

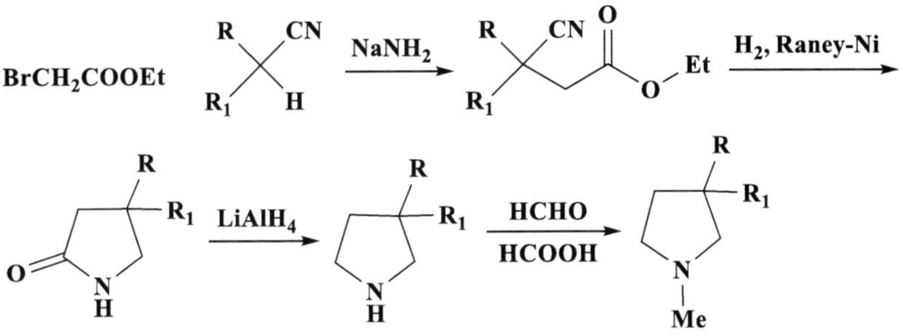

Scheme 3.2

R = p-F$_3$CC$_6$H$_4$, p-MeOC$_6$H$_4$, Ph, 2-naphthyl, p-Me$_2$NC$_6$H$_4$, p-O$_2$NC$_6$H$_4$, o-ClC$_6$H$_4$, 4-pyridyl
R$_1$ = Ph, p-Br-C$_6$H$_4$, o-ClC$_6$H$_4$

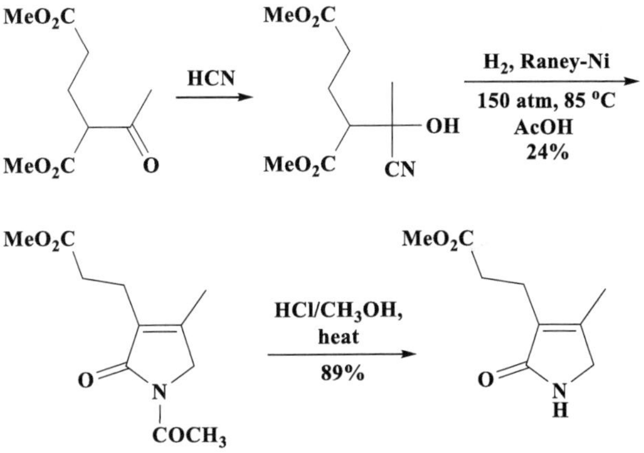

Scheme 3.3

Traditional routes to 3,3-diarylpyrrolidines suffered from the use of difficult-to-access, noncommercially available starting materials, need for expensive catalysts, and poor yields. Additionally, only a few syntheses are reported in the literature including preparation from diarylacetonitriles (Scheme 3.2) [17].

Until 1956, because of the generation of a mixture of isomeric pyrrolinones, an oxidation of $\alpha\alpha'$-free pyrroles using hydrogen peroxide was difficult when the substituents were different. An efficient, inconvenient and potentially hazardous pathway was developed by Fischer and Plieninger [18,19] for the synthesis of pyrrolinones with different substituents (Scheme 3.3). Thus, the reaction of acetylglutaric acid dimethyl ester with liquid

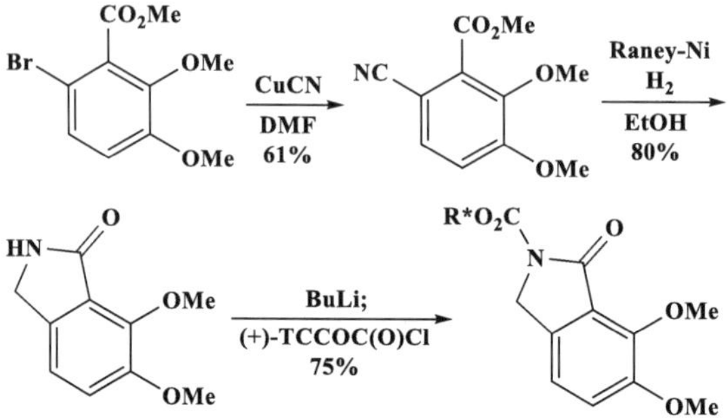
CO2Me
Br
OMe
OMe
CuCN
DMF
61%
NC
Raney-Ni
H2
EtOH
80%
HN
O
BuLi;
(+)-TCCOC(O)Cl
75%
R*O2C
N

Scheme 3.4

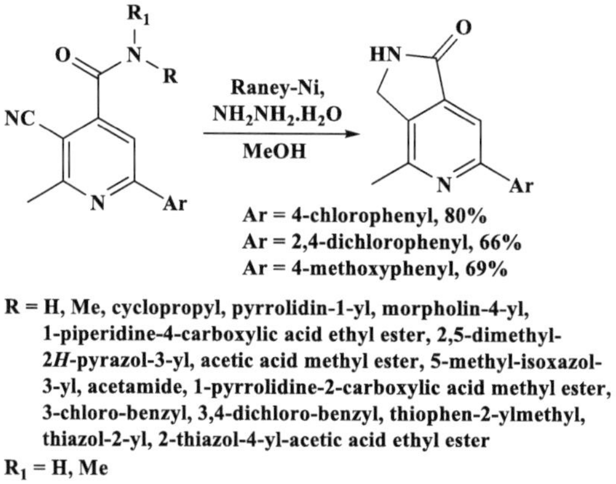
R1
O
N
R
NC
N
Ar
Raney-Ni,
NH2NH2.H2O
MeOH
HN
Ar = 4-chlorophenyl, 80%
Ar = 2,4-dichlorophenyl, 66%
Ar = 4-methoxyphenyl, 69%
R = H, Me, cyclopropyl, pyrrolidin-1-yl, morpholin-4-yl, 1-piperidine-4-carboxylic acid ethyl ester, 2,5-dimethyl-2H-pyrazol-3-yl, acetic acid methyl ester, 5-methyl-isoxazol-3-yl, acetamide, 1-pyrrolidine-2-carboxylic acid methyl ester, 3-chloro-benzyl, 3,4-dichloro-benzyl, thiophen-2-ylmethyl, thiazol-2-yl, 2-thiazol-4-yl-acetic acid ethyl ester
R1 = H, Me

Scheme 3.5

hydrogen cyanide afforded cyanohydrins at high pressure and temperature, which further gave *N*-acetylpyrrolinone by hydrogenation over Raney nickel catalyst. A valuable pyrrolinone was obtained by acid-catalyzed saponification [20–23].

Comins et al. [24] reported the first asymmetric synthesis of (+)-lennoxamine. The known isoindolone was prepared from benzoate through the addition of cyanide and reduction, which led to lactamization (Scheme 3.4) [25]. At this stage, the chiral auxiliary, (+)-*trans*-2-(cumyl)cyclohexanol (TCC), was installed through an acylation with its chloroformate [26].

The tertiary amides cyclized spontaneously in the presence of Raney nickel and hydrazine hydrate to afford the pyrrolopyridines (Scheme 3.5) [27,28].

The pyrrolo[3,2-*c*]quinoline skeleton of martinelline natural products was constructed by postfunctionalization of *N*-tosyl-6-benzyloxy-4-

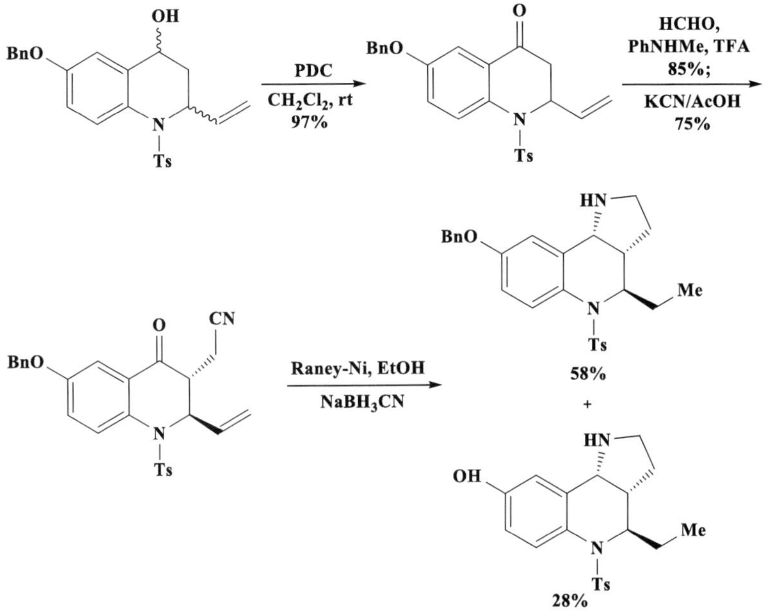

Scheme 3.6

hydroxy-2-vinyl-1,2,3,4-tetrahydroquinoline [29]. The *trans*-2,3-disubstituted tetrahydroquinolone was obtained in good yield by PDC oxidation of *N*-tosyl-6-benzyloxy-4-hydroxy-2-vinyl-1,2,3,4-tetrahydroquinoline followed by reaction with HCHO in the presence of acetic acid/potassium cyanide and *N*-methylaniline trifluoroacetate, which was finally transformed into pyrrolo[3,2-*c*]quinoline by Raney nickel reduction of the nitrile group followed by reduction of the cyclic imine with NaBH$_3$CN (Scheme 3.6) [30].

The development of a short and formal synthesis of strychnine is in progress. A simple indole precursor underwent a SmI$_2$-induced cascade reaction for the formation of an intermediate, which further constructed two new rings and three stereogenic centers (Scheme 3.7) [31,32].

Scheme 3.8 shows the synthetic pathway to pyrrolidine in which ketoacid [33] was used as a starting compound, which was further esterified to synthesize the ketoester. The TOSMIC (toluenesulfonylmethyl isocyanide) was used for the reductive cyanation of ketoester in high yield [34]. The lactams, prepared by hydrogenation of cyanoester over Raney nickel catalyst in ethanol, were further reduced with lithium aluminum hydride in tetrahydrofuran to synthesize the pyrrolidines. The *N*-alkylation of

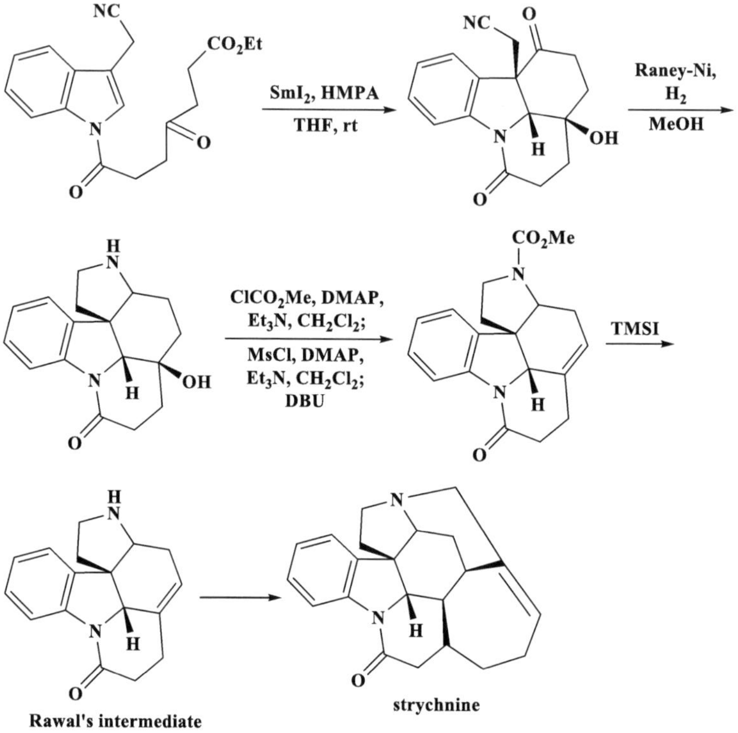

Scheme 3.7

heterocyclic secondary amines during hydrogenation has rarely been disclosed in the literature and only in cases where the reaction was performed at high temperatures [35]. The reaction mixture afforded only N-ethyl lactam after 3 h of hydrogenation at 140°C. Moreover, its partial formation was also observed at temperatures below 100°C. For the reported N-alkylation of lactams, the anticipated stability to disproportionation of the solvent (primary alcohols: ethanol, methanol) might be a logical explanation, in which, sufficiently electrophilic conditions were provided under specific reaction conditions. The nucleophilic attack by primary aminoesters, formed during hydrogenation, provided secondary aminoesters which were consecutively cyclized to N-alkylated lactams. A mixture of oxazolone and oxime was obtained when hydroxylamine was reacted with ketoester. The oxazolone was synthesized by saponification of this mixture followed by sublimation [36]. The pyrazolone was formed upon treating 2-oxo-1-adamantane carboxylic acid with hydrazine followed by sublimation [37]. The pyrazolothione was formed in 93% yield upon

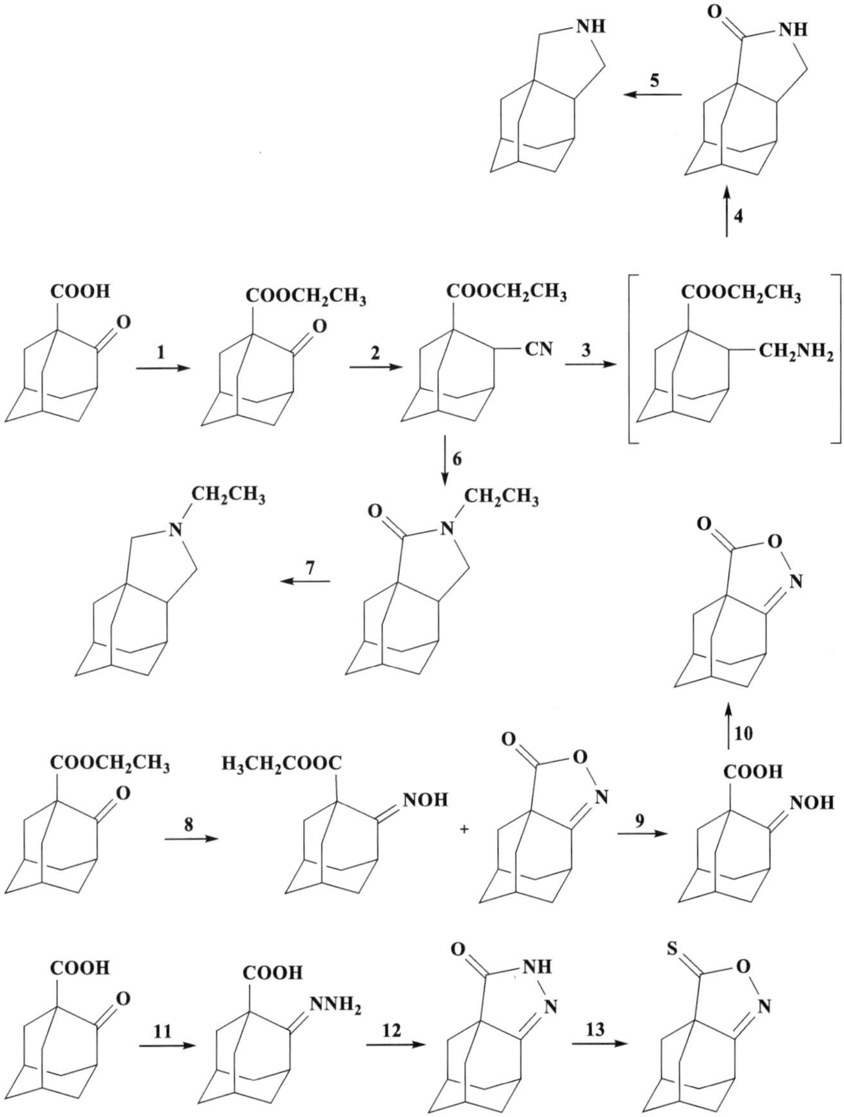

Synthesis of heterocycles from cyanide, oxime, and azo compounds using Raney nickel 87

Reagents and conditions: (1) a) SOCl$_2$, 65 °C, 15 min, b) abs. EtOH (quant.), (2) TOSMIC, abs. EtOH, DME, *t*-BuOK, 0 °C, argon, 20 °C, 30 min and 48 °C, 1 h, 74%, (3) H$_2$/Raney-Ni, MeOH, 62 psi, 60 °C, 6 h, (4) xylene, reflux, 10 h, 40%, (5) LiAlH$_4$, THF, 18 h, reflux, 70%, (6) H$_2$/Raney-Ni, EtOH, 65 psi, 140 °C, 3 h, 46%, (7) LiAlH$_4$, THF, 13 h, reflux, 92%, (8) NH$_2$OH·HCl, CH$_3$COONa·3H$_2$O, EtOH/H$_2$O (5:1), reflux, 3 h, (9) NaOH, EtOH, H$_2$O, 3.5 h, 60 °C and then conc. HCl, 97%, (10) sublimation, 10-2 mmHg, 79%, (11) NH$_2$NH$_2$, abs. EtOH, 30 min, reflux, 72%, (12) sublimation, 10-2 mmHg, 93%, (13) Lawesson's reagent, toluene, 12 h, reflux, 93%.

Scheme 3.8

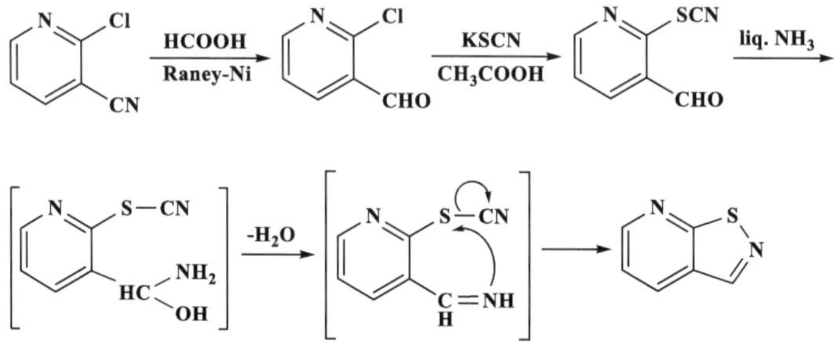

Scheme 3.9

refluxing pyrazolone in toluene with *p*-methoxyphenylthionophosphine sulfide (Lawesson's reagent) [38,39].

The reaction of *o*-thiocyano-acetyl- or *o*-thiocyano-formyl-pyridine with ammonia afforded isothiazolo[5,4-*b*]pyridine (Scheme 3.9). The 2-chloronicotinonitrile was the starting compound for the formation of isothiazolo[5,4-*b*]pyridine, which was further reduced with Raney nickel-formic acid to afford the 2-chloronicotinaldehyde. The treatment of 2-chloronicotinaldehyde with potassium thiocyanate in glacial CH_3COOH under nitrogen atmosphere provided 2-thiocyanonicotinaldehyde. The isothiazolo[5,4-*b*]pyridine was synthesized by cyclization of 2-thiocyanonicotinaldehyde at -50°C in large excess of liquid ammonia for 4 h. Three other isomeric isothiazolopyridines were formed through the disubstituted pyridines [40].

3.2.2 Synthesis of six-membered heterocycles from cyanide compounds

A convenient formation of substituted piperidine-2-ones was developed by Singh et al. [41] from Baylis–Hillman acetates through a key step, that is, reductive cyclization. The corresponding products, obtained by 1,4-diazabicyclo[2.2.2]octane-mediated bimolecular nucleophilic substitution reaction of ethyl cyanoacetate with Baylis–Hillman acetates, were easily hydrogenated with Raney-Ni for the formation of piperidine-2-ones in moderate yields. The synthesis of substituted piperidine-2,6-diones occurred through the nitrile hydrolysis and subsequent cyclization (Scheme 3.10). The highly substituted 3-methylene-2-pyridones were prepared by a similar strategy [42,43].

The first successful synthesis of a bridged lactam was described by Albertson [44] (Scheme 3.11) (curiously, regarding bridged amides this pio-

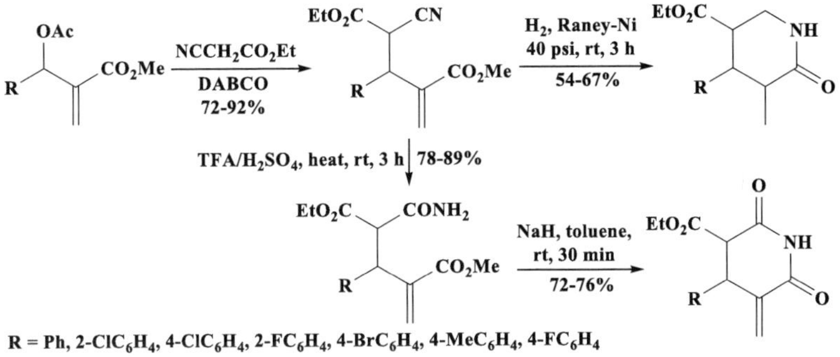

Synthesis of heterocycles from cyanide, oxime, and azo compounds using Raney nickel   89

Scheme 3.10

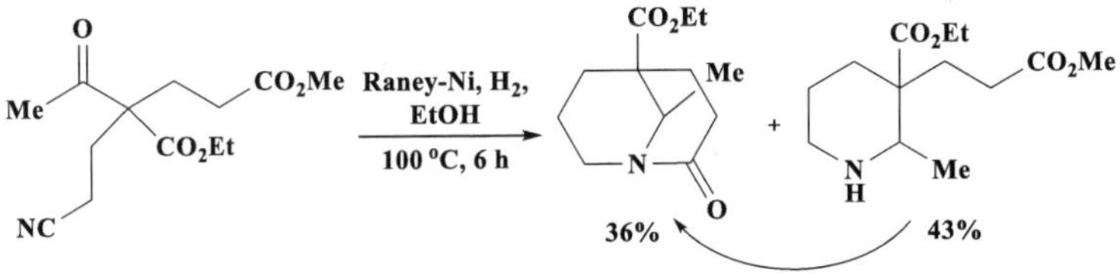

R = Ph, 2-ClC$_6$H$_4$, 4-ClC$_6$H$_4$, 2-FC$_6$H$_4$, 4-BrC$_6$H$_4$, 4-MeC$_6$H$_4$, 4-FC$_6$H$_4$

Scheme 3.11

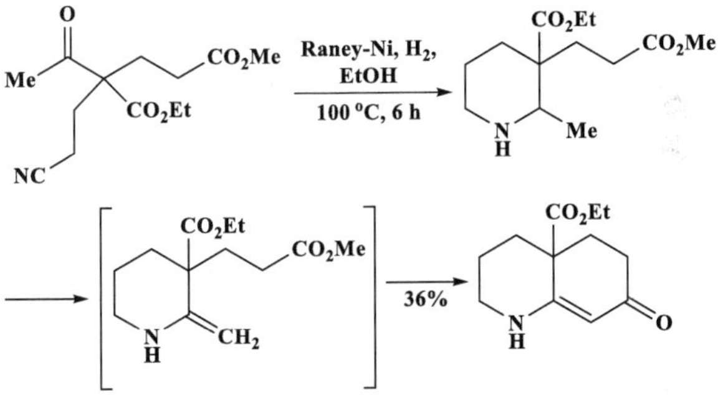

Scheme 3.12

neering example has been regularly omitted in the literature). The stability of amides with C=O bond placed on a 3-carbon bridge was much higher than when the C=O bond was located at a 2-carbon bridge as suggested by the feasibility of Albertson's synthesis [44]. Badger et al. [45] reported the first formation of 1-azabicyclo[3.3.1]nonan-2-one derivative (Scheme 3.12). An intramolecular condensation of cyano diester to bridged lactam was carried out during studies on hydrogenation reactions in the presence of copper chromite catalyst followed by reduction under reaction conditions to afford the 1-azabicyclo[3.3.1]nonane in 39% yield. Simultaneously, two

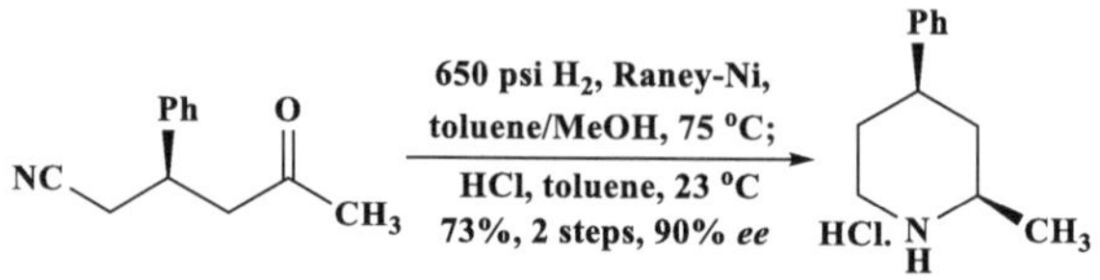

Scheme 3.13

compounds, one of which was bridged lactam, were prepared by Albertson [44] via hydrogenation of another cyano ester over Raney nickel catalyst. However, Albertson [46] reinvestigated this reaction and reported that the compound originally proposed as lactam was more consistent with the bicyclic enaminone [47].

The use of cyano derivative as a precursor to several cyclic compounds of interest was allowed by the versatility of nitrile functional group. Thus, the *cis*-2,4-disubstituted piperidine was prepared as a single diastereomer by Raney nickel-catalyzed hydrogenation, the relative stereochemistry of which was confirmed by X-ray crystallography (Scheme 3.13). Earlier, such products have been synthesized in enantioenriched form by alkylation of chiral bicyclic lactams [48]; this new pathway provided a versatile procedure for the synthesis of this class of compounds by an asymmetric catalysis [49].

An acetylation of Baylis–Hillman adducts afforded starting compounds. Baylis–Hillman adducts were synthesized from substituted benzaldehydes following the literature procedure [50,51]. The substituted 1,5-dipentanoates were obtained in excellent yields as diastereoisomeric mixture in the presence of 1,4-diazabicyclo[2.2.2]octane (DABCO) by nucleophilic substitution of ethyl cyanoacetate with acetates in a tetrahydrofuran-H_2O system following a standard procedure [52]. The 5-methyl-6-oxo-4-aryl piperidine-3-carboxylic acid ethyl esters were formed as mixtures of diastereoisomers in 54–67% yield by reduction of substituted 1,5-dipentanoates in the presence of Raney nickel under hydrogenation conditions. In principle, an oxidation of these piperidinones provided pyridine-2-one derivatives with a suitable reagent. Accordingly, the aromatization of pyridine-2-ones was examined in the presence of DDQ (2,3-dichloro-5,6-dicyanobenzoquinone) [53]. However, an oxidation did not occur and only starting compound was obtained. At this stage, the desired pyridinone was formed easily via an oxidation if 5-methyl-6-oxo-4-aryl piperidine-3-carboxylic acid ethyl esters were transformed to its chloro-derivative with phosphorusoxy trichloride [54]. Disappointingly, the chlorination with phosphorusoxy trichloride under several conditions failed to synthesize the chloro derivative. Subsequently, the chlorination of 5-methyl-6-oxo-4-aryl piperidine-3-carboxylic acid ethyl esters was attempted with a mixture of phosphorus pentachloride and phosphorusoxy trichloride. Interestingly, a less polar product was prepared by this reaction, whose structure was established on the basis of spectroscopic analysis. In principle, it acted as an appropriate substrate for dehydrohalogenation followed by an oxidation because of the tertiary nature of the chloro group. It was delightful to note that the reaction in DBU afforded good yields. The structure of these compounds, that is, 5-methyl-4-oxo-6-aryl-3-azabicyclo[3.1.0]hexane-1-carboxylates was determined by spectroscopic evidences [55]. The hydrogen atom on the carbon having alkoxycarbonyl group being more acidic was involved in the elimination of chloride ion. The transformation of cyano group to an amide led to the formation of an intermediate, which underwent an intramolecular cyclization to afford the 3-methylene piperidine-2,6-dione. An interesting ferric chloride-mediated highly efficient formation of 1,2-dihydro-2-oxo-3-pyridine-carboxylate starting from cyano ethyl acetate and enones was explained by Wang and coworkers [56]. The 3-methylene-4-substituted phenyl piperidine-2,6-diones were obtained in 60–65% yield by treating 1,5-dipentanoate derivatives with $FeCl_3 \cdot 6H_2O$ (3 eq.) in propionic acid at reflux for 2 h. However, it was noticed that the carboxylate group was removed and dehydrogenation not occurred during the reaction. The 1,5-dipentanoate derivatives were hydrolyzed in the presence of trifluoroacetic acid/sulfuric acid mixture for the formation of amides in good yields to examine the scope of substrates for the formation of carboxylate group containing piperidine-2-one derivative. The desired 5-methylene-2,6-dioxo-4-phenyl piperidine-3-carboxylic acid ethyl esters were generated in good yields by treating amides with sodium hydride at rt. Interestingly, the 5-methylene-2,6-dioxo-4-phenyl piperidine-3-carboxylic acid ethyl esters were obtained in high stereoselectivity in favor of the *trans*-isomer. The formation of 3,5-dimethylene-4-phenylpiperidine-2,6-dione and monoalkylidene glutarimide was reported by Lee and coworkers [57] by hydrolysis of cyano group with H_2SO_4 in MeOH followed by cyclization in the presence of $NaHCO_3$. However, a complex mixture of products was obtained by $NaHCO_3$-mediated cyclization of substrates amides. The 3-methylene-piperidine-2-ones were obtained but in low yield (Scheme 3.14) [41].

The formation of clavolonine, a hydroxylated *Lycopodium* alkaloid, was reported by Evans and Scheerer [58] (Scheme 3.15). An advanced diketone intermediate was generated in 11 steps using chiral oxazolidinone chemistry. The first ring of natural product was constructed by cyclization

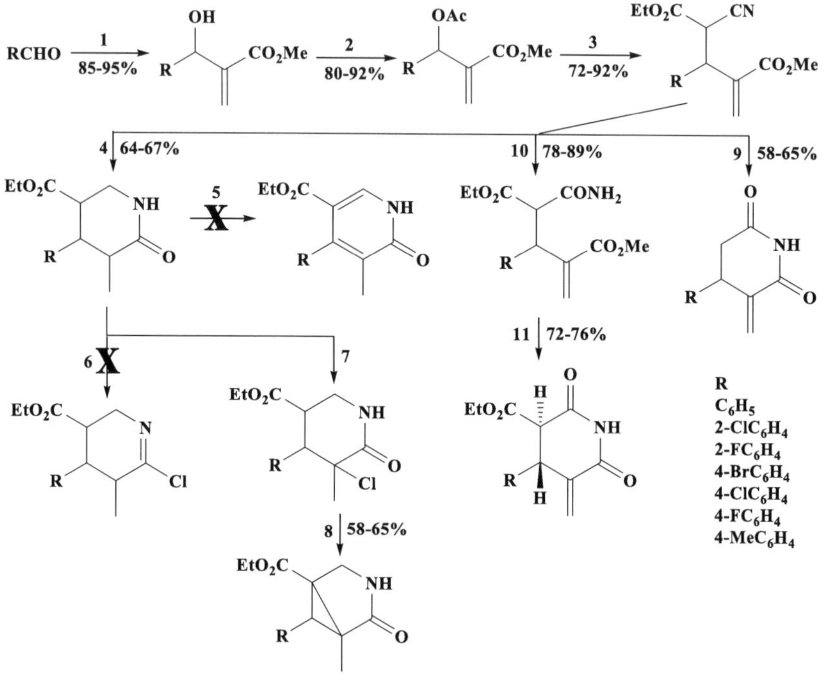

Reagents and conditions: (1) CH$_2$=CHCO$_2$Me, DABCO, rt, 2-5 d, (2) AcCl, Py, CH$_2$Cl$_2$, rt, 4-6 h, (3) CNCH$_2$CO$_2$Et, DABCO, THF/H$_2$O, rt, 2 h, (4) Raney-Ni, 40 psi H$_2$, rt, 3 h, (5) DDQ, dioxane, reflux, 24 h, (6) POCl$_3$, heat, reflux, 24 h, (7) PCl$_5$, POCl$_3$, reflux, 2 h, (8) DBU, CH$_3$CN, reflux, 14 h, (9) FeCl$_3$·6H$_2$O, propionic acid, reflux, 2 h, (10) TFA/H$_2$SO$_4$, heat, rt, 3 h, (11) NaH, toluene, rt, 30 min.

Scheme 3.14

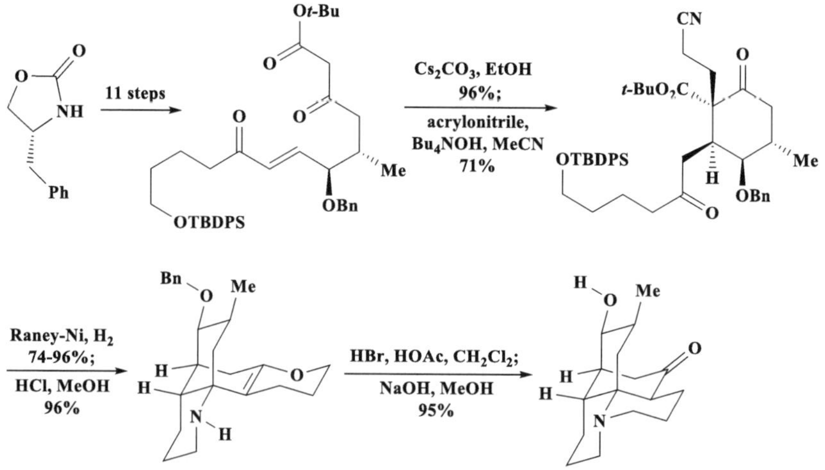

Scheme 3.15

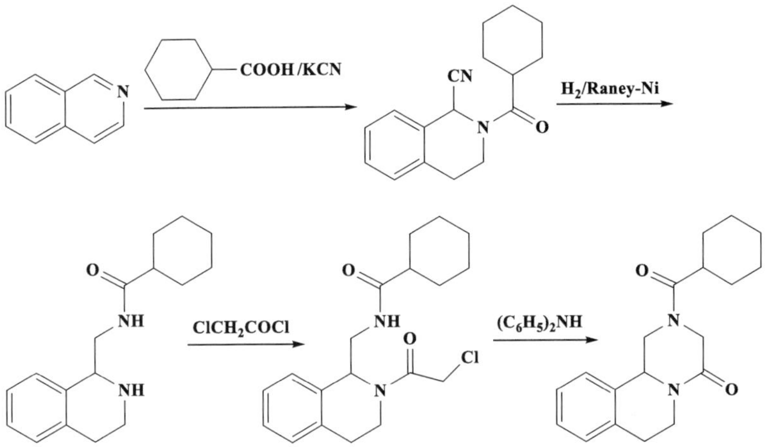

Scheme 3.16

of diketone intermediate followed by an intermolecular Michael reaction with acroylnitrile to synthesize the highly functionalized cyanoketone. The transformation of cyanoketone to cyclic imine occurred by reduction with Raney nickel. Subsequently, treatment of imine with HCl triggered a decarboxylative Mannich cascade with contaminant cyclic enol ether formation to give the tetracycle. The enol ether was treated with hydrogen bromide to liberate the ketone and form the bromide, which was promptly alkylatively cyclized onto the nitrogen. The clavolonine was generated by deprotonation of hydrogen bromide salt with sodium hydroxide.

The treatment of isoquinoline with a mixture of cyclohexanecarbonyl chloride/KCN to synthesize a dihydro derivative of isoquinoline is a way of synthesizing praziquantel. The reduction-reamidation product, that is, amide 1-(N-cyclohexylcarbonylaminomethyl)-1,2,3,4-tetrahydroisoquinoline was obtained by further reduction with H_2 over Raney nickel. The chloroacetyl derivative, formed by acylating with chloroacetic acid chloride, was heated in the presence of diethylamine for an intramolecular alkylation to afford the praziquantel (Scheme 3.16) [59].

It became apparent that problems with the deprotonation of nitrile accounted for the unsuccessful halomethylation reactions. As nitrile α-protons are known to be sufficiently acidic to be deprotonated by lithium diisopropylamide, it was thought that the synthetic problem could stem from the stability of the carbanion formed following deprotonation of the nitrile. In order to trap the apparently unstable carbanion immediately after deprotonation, a reaction was carried out in which a precooled solution of lithium diisopropylamide was added to a solution containing both the cyclopentylcarbonitrile and excess electrophile (bromochloromethane) at −78°C. This procedure successfully gave desired halomethylated product, though with a low yield (24%). The proton nuclear magnetic resonance and mass spectrometry analysis of mixture of by-products indicated polymeric species containing several chlorine atoms, a problem reported to occur in the photochemical synthesis of 2,2-dialkylethanenitrile and in the reaction of dihalomethanes with 2-arylethanenitriles [60]. A sufficient amount of desired halomethylated intermediate was obtained to proceed with the synthesis of target pyrimidoindolone probe. However, in order to prepare a library for future SAR studies of pyrimidoindolone heterocycles, it would be necessary to improve the efficiency and the yield of 2,2-dialkyl-3-halopropanenitrile. Following the successful synthesis of the alkylating agent, 2-cyclopentyl-3-chloropropanenitrile, the previously prepared ketal-protected isatin sulfonamide was successfully alkylated by bimolecular nucleophilic substitution to give the corresponding nitrile in 71% yield. The next step involved the reduction of nitrile group to a primary amine, which underwent an intramolecular dehydration/cyclization to give the amidine ring. An initial reduction attempt using catalytic Raney-Ni at atmospheric pressure for up to 2 days was unsuccessful. The nitrile reduction was ultimately achieved using an autoclave at a pressure of ∼70 psi and a very dilute solution. An autoclave was required to increase the pressure and the concentration of hydrogen gas in the reaction mixture, while a dilute solution prevented intermolecular reactions which may compete with the subsequent desired intramolecular cyclization. The amidine was isolated as the major product and an intermediate primary amine as the minor product from the hydrogenation reaction mixture. The filtration to remove the Raney-Ni catalyst and heating the reduction reaction mixture fully converted primary amine to amidine, with a total yield of 83% (Scheme 3.17) [61].

Disappointingly, efforts to selectively reduce nitrile in the presence of alkyne group, utilizing Raney nickel led to full reduction of both the alkyne and nitrile group, providing ether (Scheme 3.18) [62].

The pyrimidoindolone was prepared simultaneously with the target pyrimidoindolone probe, in five steps starting with the commercially available isatin ring. The isatin was protected as 1,3-dioxolane using 1,3-propanediol and toluenesulfonic acid in toluene to give the ketal in 50% yield. The ketal was then alkylated with chloromethyl cyclopentanecarbonitrile using *t*-BuOK as a base to give the nitrile in 14% yield with less

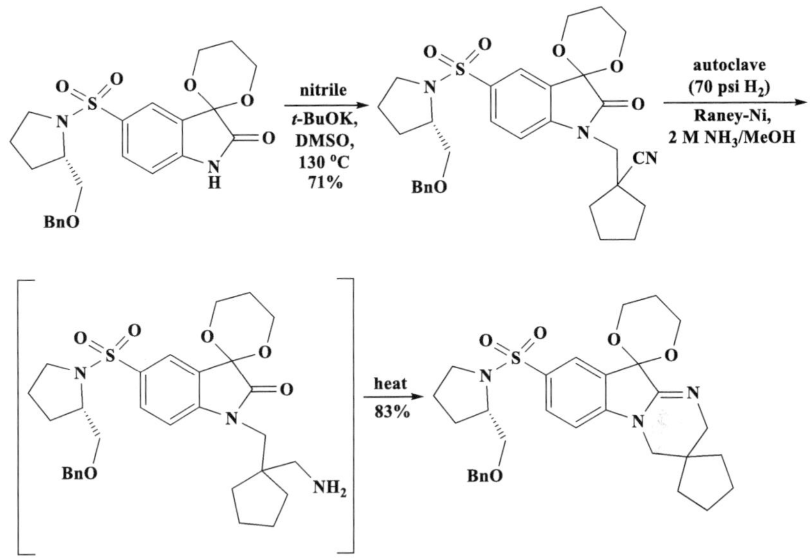

Scheme 3.17

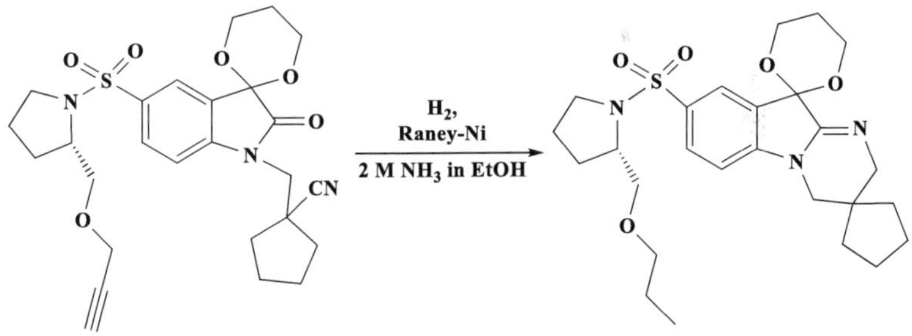

Scheme 3.18

than 80% purity. This reaction step failed to reach completion, with both the starting amide and alkylating agent isolated from the reaction mixture despite prolonged heating for up to 3 days, and using excess amounts of the base and alkylating agent. Unlike the facile amide-alkylation of sulfonamide-substituted isatin ring, alkylation of unsubstituted ring proved challenging. It was likely that the presence of an electron-withdrawing substituent on the phenyl ring, such as a sulfonamide was required to activate the ring toward an electrophilic substitution. The reduction of crude nitrile using Raney-Ni as a catalyst and methanolonic ammonia as a

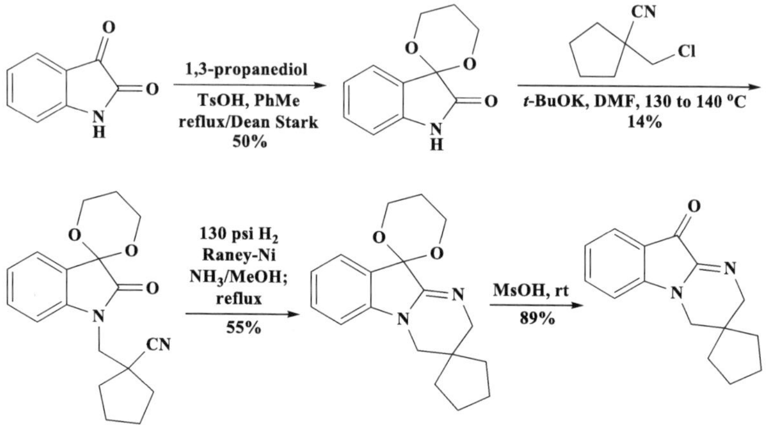

Scheme 3.19

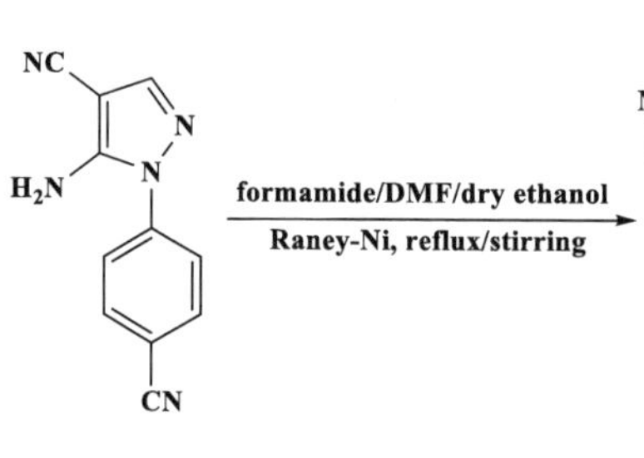

Scheme 3.20

solvent gave primary amine, which was subsequently heated to drive the cyclization/dehydration toward amidine in 55% yield. The ketal deprotection using methanesulfonic acid gave pyrimidoindolone in 89% yield (Scheme 3.19) [62].

A series of pyrazolopyrimidines was prepared to develop potent anticonvulsant agents of synthetic origin. A series of 1-(4-(aminomethyl)phenyl)-1H-pyrazolo[3,4-d]pyrimidine-4-amine, prepared by the reaction of 5-amino-1-(4-cyanophenyl)-1H-pyrazole-4-carbonitrile with formamide at 180°C using DMF as a solvent, underwent a cyclization to give an intermediate 4-(4-amino-1H-pyrazolo[3,4-d]pyrimidine-1-yl)benzonitrile, which was reduced with Raney nickel using KBH$_4$ as a reducing agent in dry EtOH (Scheme 3.20). The anticatatonic activity of test compounds was studied and found that 200 m/kg dose has protected the animals form catatonia. The presence of electron-withdrawing and electron-releasing

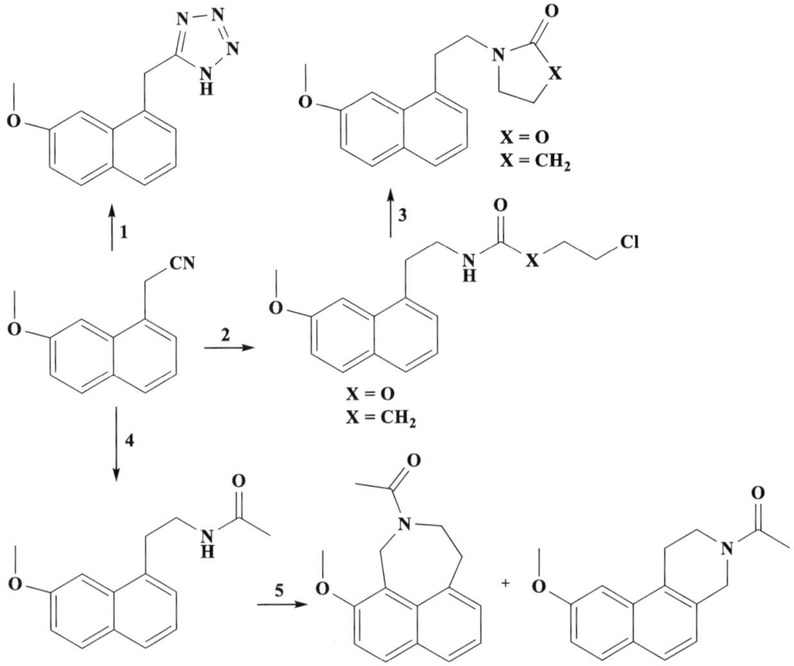

Scheme 3.21

Reagents and conditions: (1) Bu₃SnCl, NaN₃, DMF, 130 °C, (2) a) H₂, Raney-Ni, EtOH, NH₃ (gas), 50 bars, 60 °C, b) ClCO₂(CH₂)₂Cl or ClCO(CH₂)₃Cl, NEt₃, THF, rt, (3) 20% NaOH, THF, reflux, (4) H₂, Raney-Ni, (CH₃CO)₂O, 50 bars, 60 °C, 4 h, (5) DMSO, reflux.

groups has shown moderate activity when compared with that of standard drug [63].

3.2.3 Synthesis of seven-membered heterocycles from cyanide compounds

Scheme 3.21 shows the synthetic pathway to naphthalenic compounds. The tetrazole was prepared by treating 2-(7-methoxynaphth-1-yl)acetonitrile with NaN₃ in the presence of Bu₃SnCl in dimethylformamide [64]. The N-acetylated derivatives were synthesized from 2-(7-methoxynaphth-1-yl)acetonitrile by reduction of nitrile group and treatment with suitable acyl chloride. The oxazolidinone was obtained when carbamate was cyclized on heating with alkaline solution (sodium hydroxide) [65,66]. The transformation of agomelatine to cyclic compounds occurred by heating in a large excess of dimethylsulfoxide [67,68].

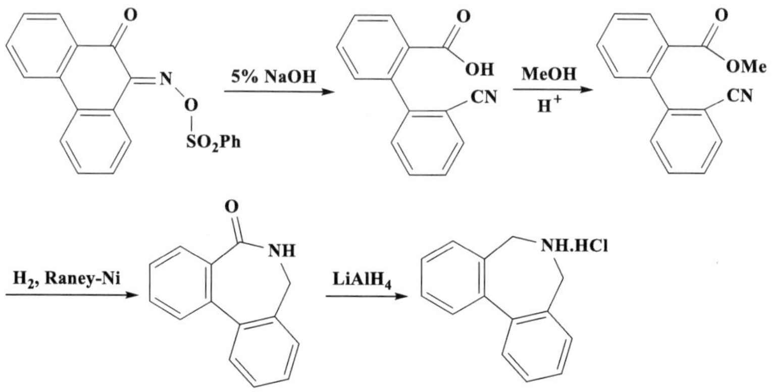

Scheme 3.22

Continuing on the investigation of benzo- and heteroazepines [69,70], an undeveloped synthon 6,7-dihydro-5H-dibenz[c,e]azepine was encountered. A photochemical reaction of N-benzyl-N-(2-iodobenzyl)amine or the reaction of o,o'-dibromoethyldiphenyl compound and NH$_3$ afforded 6,7-dihydro-5H-dibenz[c,e]azepine in very low yield (not more than 20% yield) [71a,b]. A sympatholytic activity was shown by derivatives with substituents in 6-position (especially 6-allyl) (azapetine). This is shown by a pharmacological examination of azepine and its derivatives. The azapetine and its derivatives exhibit a selective α-adreno blocking action [72]. This acted as a motivation in search of a suitable pathway to synthesize the azepine and its derivatives. The 6,7-dihydro-5H-dibenz[c,e]azepin-7-one was obtained in 60% yield by a reductive cyclization of methyl 2-(2-cyanophenyl)benzoate with H$_2$ over Raney nickel. The lactam was easily soluble in ether and provided a good yield of azepine, which was reduced with LiAlH$_4$ in ether (Scheme 3.22) [73].

An efficient and stereocontrolled formation of tryptophan- and phenylalanine-derived 5-phenyl-1,4-benzodiazepines was disclosed by Herrero et al. [74]. The Strecker reaction of N-protected amino aldehydes with TMSCN and 2-aminobenzophenone provided amino nitrile. This reaction involved a one-pot cyano reduction and reductive cyclization of appropriate amino nitrile. The 2,4-disubstituted 5-phenyl-2,3,4,5-tetrahydro-1H-1,4-benzodiazepine was prepared by subsequent reduction of 2,3-dihydro-1H-1,4-benzodiazepines followed by regioselective alkylation or acylation at the 4-position (Scheme 3.23) [75].

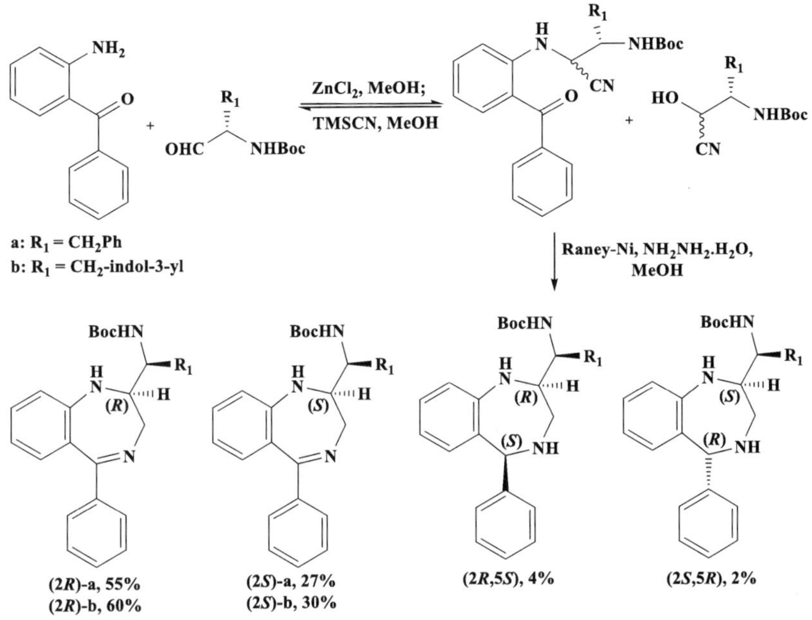

Scheme 3.23

The *N*,*N*-disubstituted amide bond was introduced in the dipeptide sequence to support the formation of diazepane (Scheme 3.24). Due to this, the reaction between acid and *N*-benzylglycine methyl ester occurred using oxalyl chloride to afford the desired amide in 77% yield. The desired amine was synthesized by treating amide with Raney nickel and H$_2$ in NH$_3$/MeOH but instead it afforded pyrrole compound. The treatment of nitrile with triethylamine also afforded pyrrole compound even in the absence of Raney nickel. The desired product was obtained in 63% yield when it was attempted to crystallize enamine from CH$_2$Cl$_2$/heptanes [76].

The synthesis of diazepane was examined for a more hindered precursor. The reaction of acid with *N*-benzylalanine methyl ester occurred in the presence of oxalyl chloride to afford the intermediate compound as a mixture of diastereoisomers (Scheme 3.25). It was reported that spontaneous cyclization occurred, which led to the formation of diazepane as a mixture of two diastereoisomers in 98% yield on reduction of the nitrile group [76].

Since high biological activity is possessed by pyrrolo[2,1-*c*][1,4]benzodiazepines [77–79], the spiro-fused diazepane was synthesized from proline and acid. The diastereoisomers were obtained in 93% yield

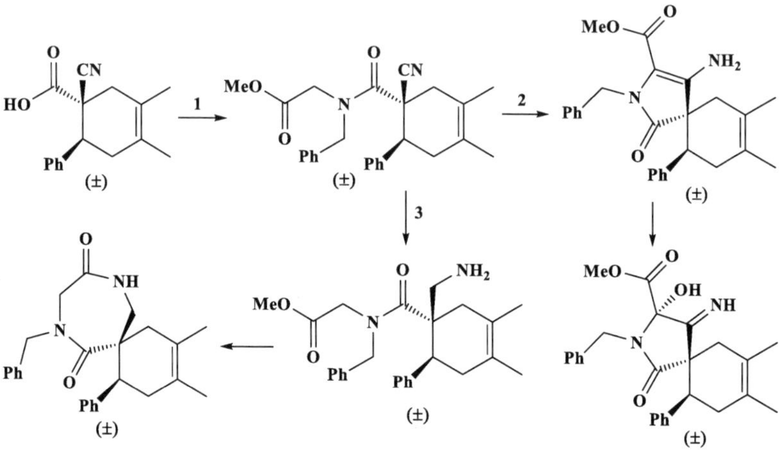

Reagents and conditions: (1) (COCl)$_2$, DMF, THF, 0 °C, 2 h, then Et$_3$N, Bn-Gly-OMe, CH$_2$Cl$_2$, rt, 20 h, 77%, (2) Et$_3$N, MeOH, rt, 20 h, 100%, (3) Raney-Ni, 50 bar H$_2$, 80 °C, 17 h, 72%.

Scheme 3.24

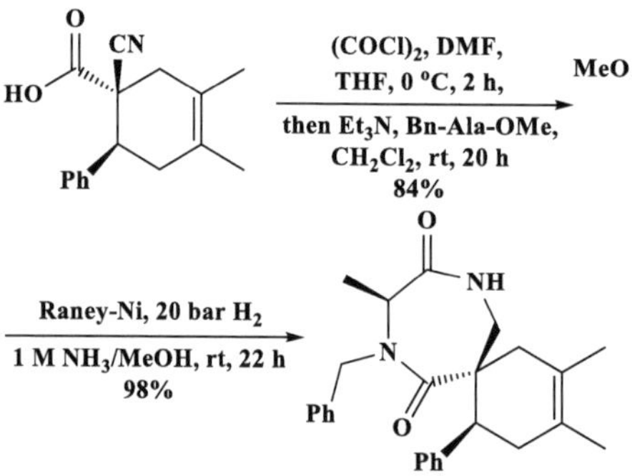

Scheme 3.25

by the reaction of H-Pro-O*t*-Bu with acid (Scheme 3.26). Raney nickel was used for the hydrogenation of nitrile group of diastereoisomers. The diastereoisomers were reduced to synthesize the amine and diazepane. The amine was heated in the presence of TEA in MeOH for the formation of diazepane. The reaction completed after 3 days with the formation of products in 3.5:1 ratio, respectively. Under nonbasic conditions, the cyclization of amine was performed to avoid the diazepane formation. The diazepane was formed in 88% yield by refluxing a solution of amine in PhMe for 5 days. The X-ray crystallography was used for the determination of the absolute stereochemistry of diazepanes [76].

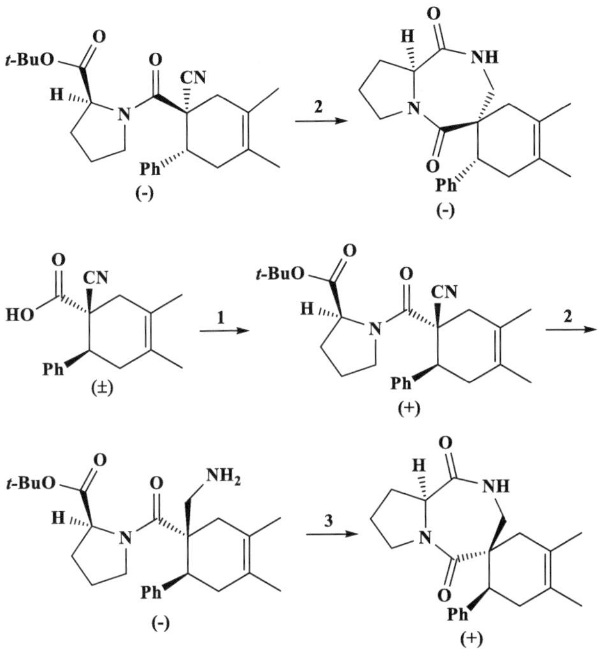

Reagents and conditions: (1) (COCl)$_2$, DMF, THF, 0 °C, 2 h, then Et$_3$N, H-Pro-Ot-Bu, CH$_2$Cl$_2$, rt, 17 h, 93%, (2) Raney-Ni, 40 bar H$_2$, 1 M NH$_3$/MeOH, rt, 17 h, (3) toluene, reflux, 5 d, 88%.

Scheme 3.26

3.3 Synthesis of heterocycles from oxime compounds

3.3.1 Synthesis of five-membered heterocycles from oxime compounds

The triol was obtained in 56% yield (based on the reactive enantiomer) by condensation of (+)-3,3-diethoxy-2-hydroxypropanal with hydroxypyruvate using transketolase. Further, the triol was silylated using *t*-butyldimethylsilyl trifluoromethanesulfonate and triethylamine (74%) and then was treated with potassium bicarbonate and hydroxylamine hydrochloride in methanol to provide the oxime in 82% yield. The reduction of oxime with Raney nickel proved capricious providing diastereomeric mixture of amines in up to 65% yield. A major obstacle was the unreliability of the oxime reduction for the successful formation of nectrisine and the availability of material was critically limited for further studies. The reaction of iodotrimethylsilane in anhydrous dichloromethane afforded a mixture of diastereomeric amines, which were easily cyclized (97%) to provide a

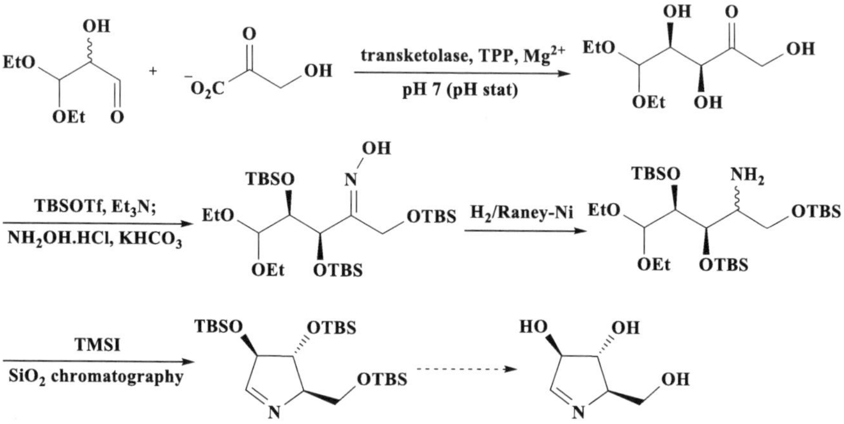

Scheme 3.27

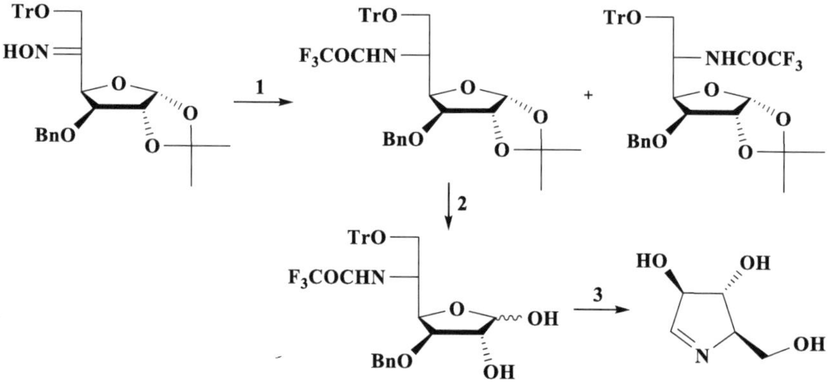

Scheme 3.28

Reagents and conditions: (1) a) H$_2$, Raney-Ni, NH$_4$OH, CH$_3$OH, b) TFAA, NEt$_3$, CH$_2$Cl$_2$, (2) 75% aq. TFA, 82%, (3) a) NaIO$_4$, THF, H$_2$O, b) H$_2$, Pd black, HCO$_2$H, CH$_3$OH, c) 1 N NaOH.

mixture of cyclic imines in 3:2 ratio from which the major diastereomer was isolated which has the stereochemistry found in nectrisine. Disappointingly, the reaction of protected imine under a variety of desilylation conditions (e.g., tetrabutylammonium fluoride; acetic acid/water/tetrahydrofuran; fluoride resin; hydrogen fluoride/acetonitrile) failed to provide a pure sample of nectrisine (Scheme 3.27) [80].

The nectrisine was prepared stereospecifically from diacetone D-glucose (Scheme 3.28) [81,82]. The catalytic reduction of starting material followed by acylation with trifluoroacetic anhydride provided trifluoroacetyl amides

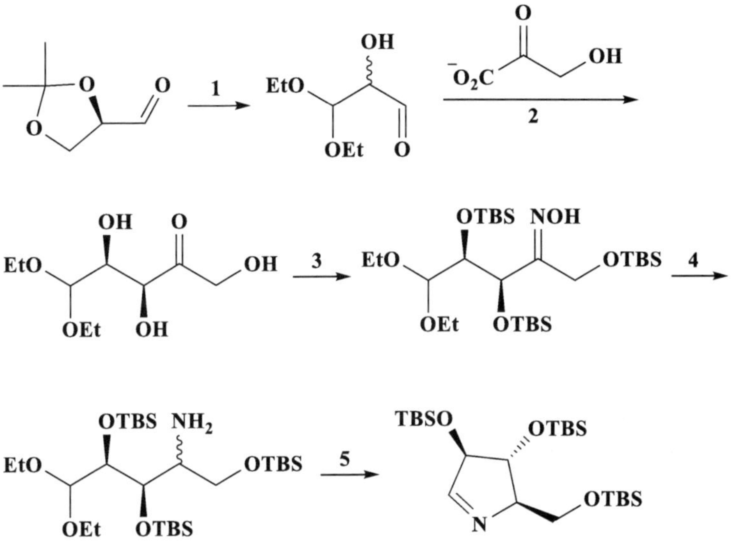

Reagents and conditions: (1) a) TEMPO, EtOAc, toluene, 0 °C, Bn(CH$_3$)$_3$NCl, NaBr, aq. NaHCO$_3$, b) 1.1 M NaOCl, 15 h, 17%, (2) transketolase, thiamine pyrophosphate Mg^{2+}, pH 7 (pH stat), (3) a) TBSOTf, NEt$_3$, b) NH$_2$OH.HCl, KHCO$_3$, CH$_3$OH, (4) H$_2$, Raney-Ni, (5) a) TMSI, b) SiO$_2$ chromatography.

Scheme 3.29

in 82% and 13% yield. The compound, obtained by the removal of *i*-propylidene group from trifluoroacetyl amides, was further oxidized with sodium periodate followed by removal of benzyl group and subsequent deacylation of formyl and trifluoroacetyl groups to synthesize the nectrisine.

A protected nectrisine was prepared from D-glyceraldehyde acetonide (Scheme 3.29) [80]. The D-glyceraldehyde acetonide was further transformed to 3,3-diethoxy-2-hydroxypropanal, which underwent a transketolase-mediated condensation with hydroxypyruvate for the formation of triol in 56% yield. The triethylamine (74%) and *t*-butyldimethylsilyl trifluoromethanesulfonate were utilized for the silylation of triol, which was further treated with hydroxylamine to synthesize the oxime in 82% yield. The diastereomeric mixture of amines (65%) was obtained by the reduction of oxime with Raney nickel. Then, the oxime was cyclized upon treatment with iodotrimethylsilane to provide a 3:2 mixture of cyclic imines from which the major diastereomer was isolated.

The 2-oxo-1-adamantanacetic acid [83–85] was utilized as a starting compound for the formation of pyrrolidine (Scheme 3.30). The acid was esterified to ethyl ester. The oxime ester was obtained when ethyl ester was

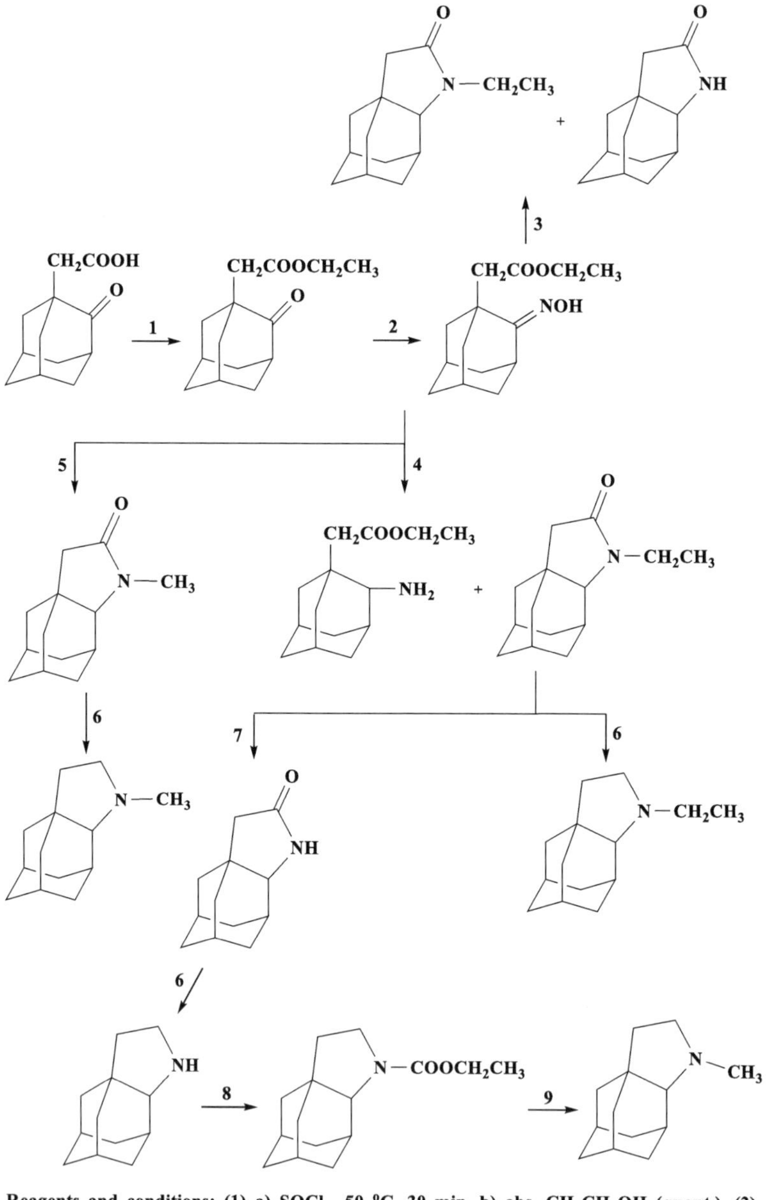

Reagents and conditions: (1) a) SOCl$_2$, 50 °C, 30 min, b) abs. CH$_3$CH$_2$OH (quant.), (2) NH$_2$OH·HCl, CH$_3$COONa·3H$_2$O, CH$_3$CH$_2$OH/H$_2$O (5:1), reflux, 1 h, 97%, (3) a) H$_2$/Raney-Ni, EtOH, 55 psi, 120 °C, 10 h, b) xylene, reflux, 12 h, (4) H$_2$/Raney-Ni, EtOH, 55 psi, 100 °C, 3 h, (5) a) H$_2$/Raney-Ni, MeOH, 55 psi, 200 °C, 4 h, b) xylene, reflux, 20 h, 64%, (6) LiAlH$_4$, THF, 5 h, reflux, 94–96%, (7) xylene, reflux, 12 h, (quant.), (8) Et$_3$N, ClCOOC$_2$H$_5$, ether, 24 h, 25 °C, 96%, (9) LiAlH$_4$, THF, 24 h, 25 °C, 93%.

Scheme 3.30

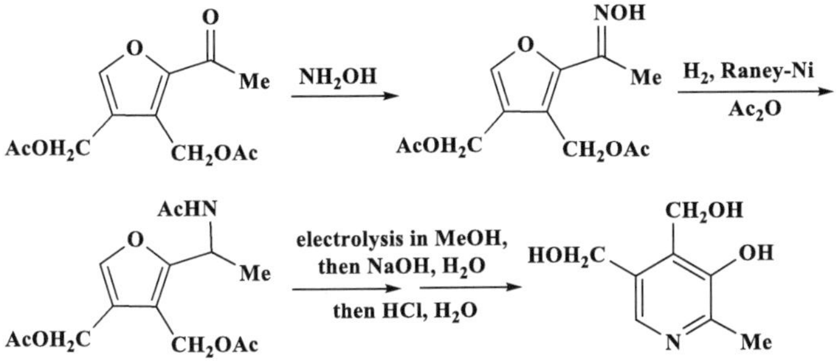

Scheme 3.31

reacted with NH$_2$OH·HCl in the presence of CH$_3$COONa. A mixture of lactams was obtained by catalytic hydrogenation of oxime ester over Raney nickel under different reaction conditions. The pyrrolidines were synthesized by the reduction of lactams with lithium aluminum hydride in tetrahydrofuran [86]. The N-methyl derivative was obtained by N-acylation of pyrrolidine followed by reduction of the carbamate intermediate with lithium aluminum hydride [39].

3.3.2 Synthesis of six-membered heterocycles from oxime compounds

Clauson-Kaas and Nedenskov [87] synthesized pyridoxine in 76% overall yield. They synthesized 2-acetyl-3,4-bis(acetoxymethyl)furan applying the same method as Williams et al. [88] reported earlier, but then the acyl group was reacted with hydroxylamine hydrate to synthesize the oxime. The oxime was reduced with H$_2$ using a Raney-Ni catalyst followed by acetylation to give the 2-(acetamidomethyl)-3,4-bis(acetoxymethanol)furan, which provided pyridoxine via a three-step synthesis where none of the intermediates was isolated (Scheme 3.31).

The amide was cyclized with CSA (camphor sulfonic acid)-catalyzed reaction, which provided desired enamide in 85% yield (Scheme 3.32). The oxime was obtained when enamide was reacted with NOCl and then reduced in situ with NaBH$_3$CN. This oxime was reduced under high pressure hydrogenation with Raney nickel. Only one isomer with correct stereochemistry was obtained when hydrogenation occurred from the less hindered face of the oxime. With the amine in hand, the Pictet-Spengler cyclization was performed with benzyloxyacetaldehyde followed

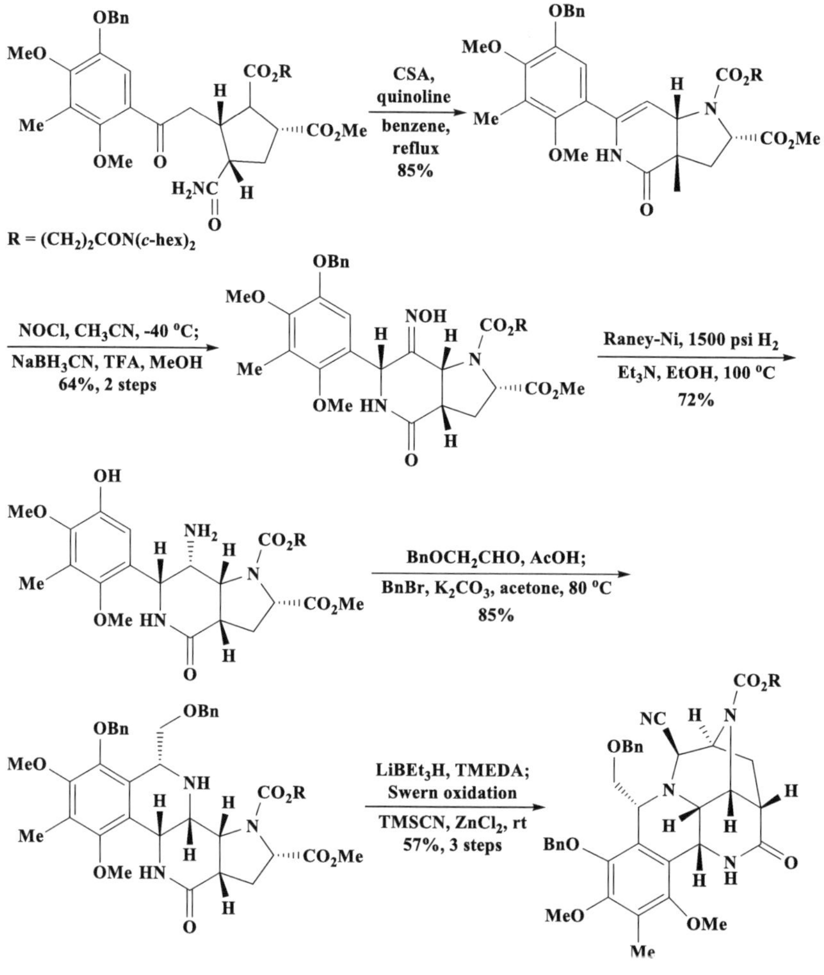

Scheme 3.32

by protection of phenol as benzyl ether. A mixture of aminals was obtained upon reduction of methyl ester with LiBEt$_3$H and then Swern oxidation of alcohol to aldehyde resulting in the spontaneous ring-closure. These unstable aminals were trapped as a single aminonitrile in the presence of ZnCl$_2$ and TMSCN [89].

Ishikawa et al. [90,91] cyclized oximes in the presence of Raney nickel and sodium hydroxide (Scheme 3.33) [92].

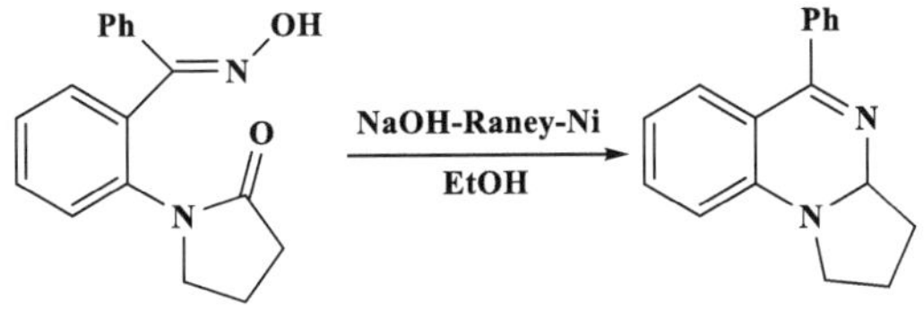

Scheme 3.33

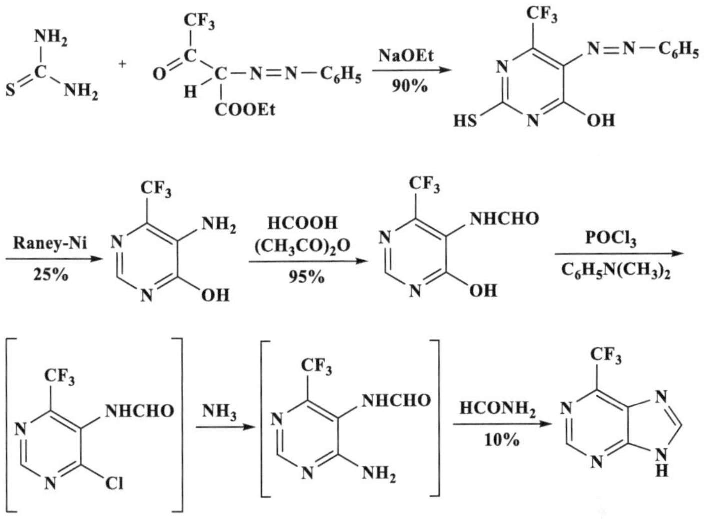

Scheme 3.34

3.4 Synthesis of heterocycles from azo compounds

A more fruitful approach to purine [93] proceeded through an ethylphenylazotrifluoroacetoacetate, which on condensation with thiourea afforded pyrimidine in excellent yields. The 5-amino-4-hydroxy-6-trifluoromethylpyrimidine was prepared by concomitant desulfurization and hydrogenolysis with Raney nickel. For the replacement of 4-hydroxyl by chlorine atom with phosphorusoxy trichloride, and dimethylaniline, transformation to N-formyl derivative was necessary, which, further provided amine on treating with cold ethanolic ammonia. The intermediates were not obtained in pure form; however, refluxing of an intermediate in anhydrous formamide afforded desired purine in 10% overall yield (Scheme 3.34). The 2-amino-4-hydroxy-5-phenylazo-6-trifluoromethylpyrimidine was prepared directly by smooth condensation of guanidine with ethylphenylazotrifluoroacetoacetate. The 4-amino derivative was obtained in good yield by the

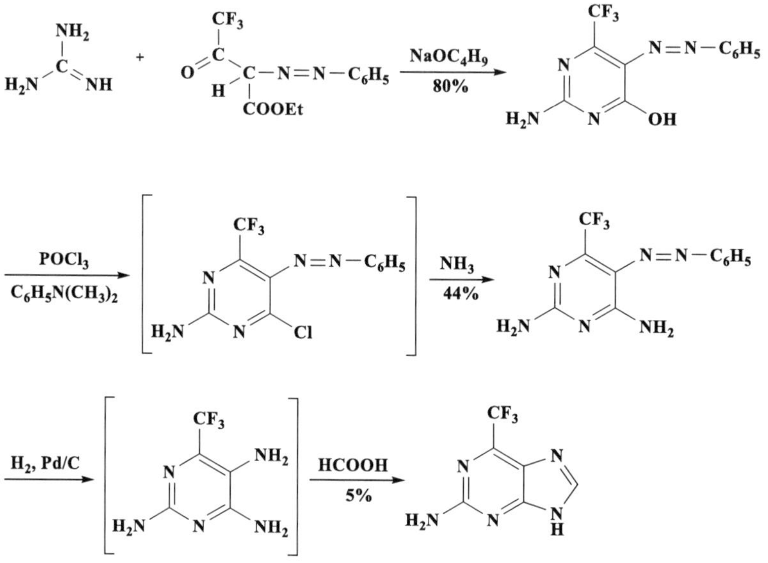

Scheme 3.35

replacement of hydroxyl with chlorine atom and subsequent amination. However, as 2,4,5-triamino-6-trifluoromethylpyrimidine was unusually unstable when exposed to air, its synthesis from 4-amino derivative with Raney-Ni was discontinued in favor of the milder hydrogenolysis with Pd/charcoal at rt. The triamine was heated in anhydrous formic acid to provide the 2-amino-6-trifluoromethylpurine in low yield (Scheme 3.35).

Reisman et al. [94,95] reported the synthesis of (+)-welwitindolinone A. The reaction started with a diastereo- and regioselective [2+2]-cycloaddition of ketene and cyclohexadiene acetonide obtained from acyl chloride to provide the adduct (Scheme 3.36). The o-metallated aniline equivalent was utilized for the introduction of required arylamine moiety. Raney-Ni was used for the reduction of triazene, and the amine and hydroxyl functionalities in the formed hydroxylamine were protected as a cyclic carbamate using 4-nitrophenyl chloroformate. The diol in cyclic carbamate was unmasked by acid-assisted process followed by selective oxidation of allylic alcohol under mild oxidative conditions to afford the secondary alcohol. The triisopropylsilyl ether, obtained from secondary alcohol in one-step, was transformed to enol triflate using N-phenyltriflimide and L-selectride. Many steps were needed for the central cyclohexene ring functionalization to provide the advanced intermediate ultimately.

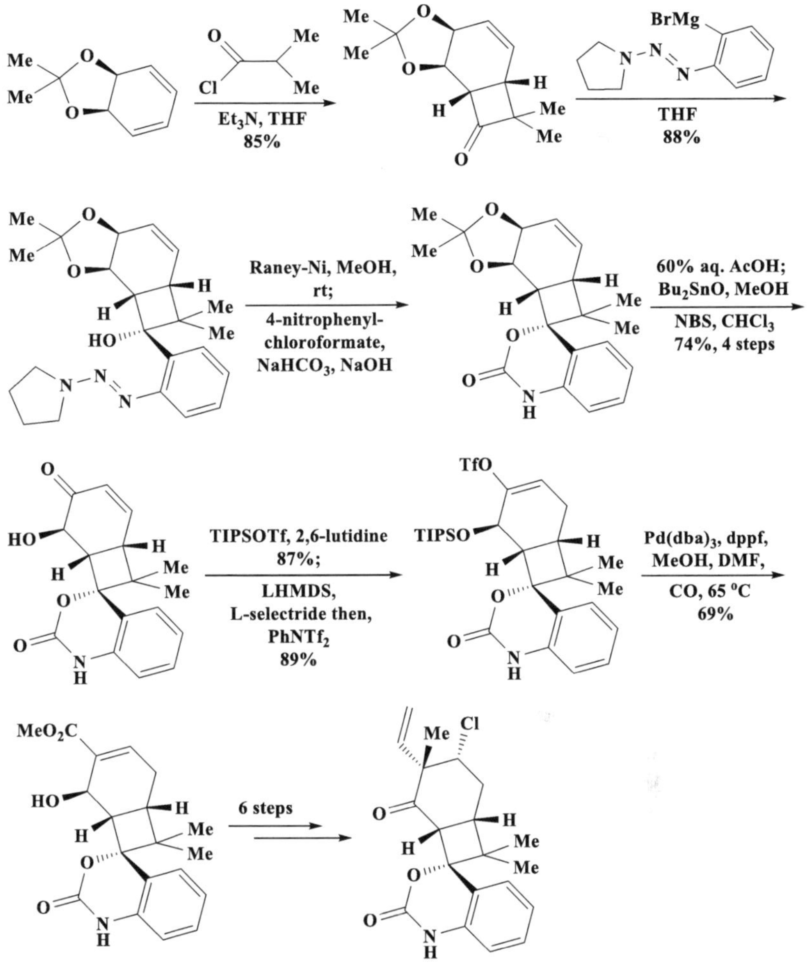

Synthesis of heterocycles from cyanide, oxime, and azo compounds using Raney nickel 109

Scheme 3.36

An O–N acyl migration and the formation of amino functionality occurred by catalytic hydrogenation of steroidal α-azido ketones with β-acyloxy groups. Thus, the N-acylated products were synthesized by hydrogenation of 17α-azido-3β,16α-dihydroxy-5α-pregnane-11,20-dione esters in the presence of Adams platinum oxide or Raney-Ni catalysts, which can be further converted into oxazolines. The O–N acyl migration and the acylation of 3β-hydroxy group took place during reduction in the case of valeryl derivative (R = Bu) (Scheme 3.37) [96]. Apparently, the O–N acyl migration required the amino and acyloxy groups in *cis*-position, therefore,

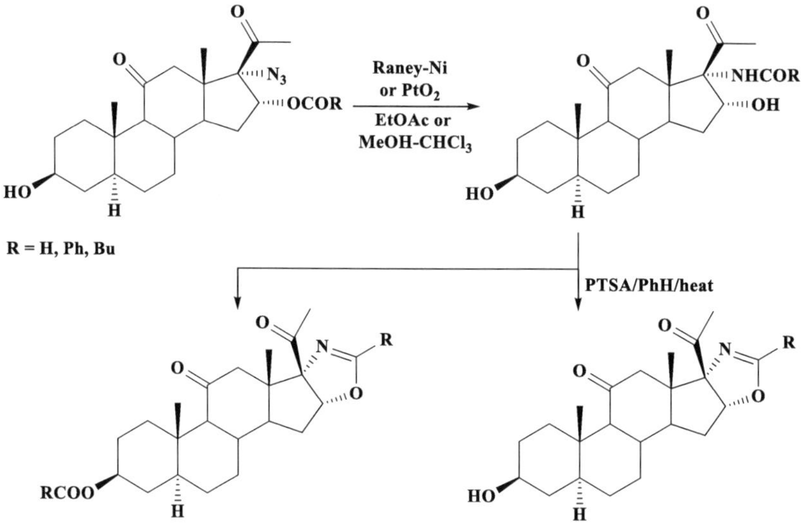

Scheme 3.37

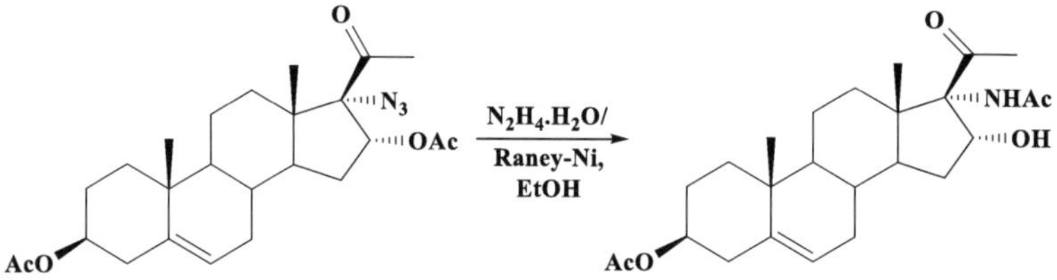

R = H, Ph, Bu

Scheme 3.38

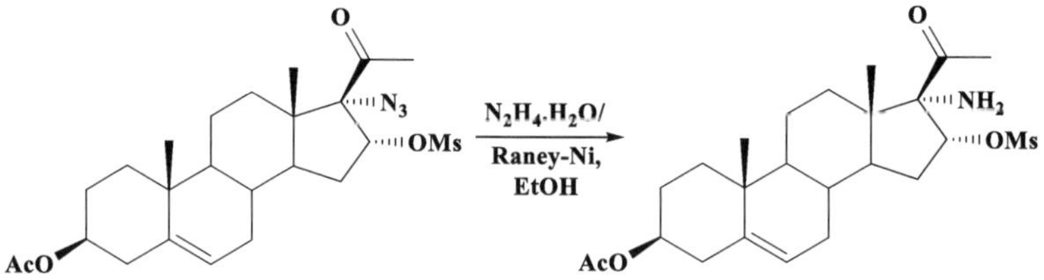

Scheme 3.39

this secondary reaction depends on the stereochemistry of the substrate. The 16α-acetoxy-17α-azidopregnenon was reduced with hydrazine hydrate using Raney nickel to prepare the N-acetyl product, which was further transformed to oxazoline under acidic conditions (Scheme 3.38). The "regular" reduction product was obtained by the reduction of 16α-mesyloxy-17α-azidopregnenon (Scheme 3.39), while an intramolecular nucleophilic

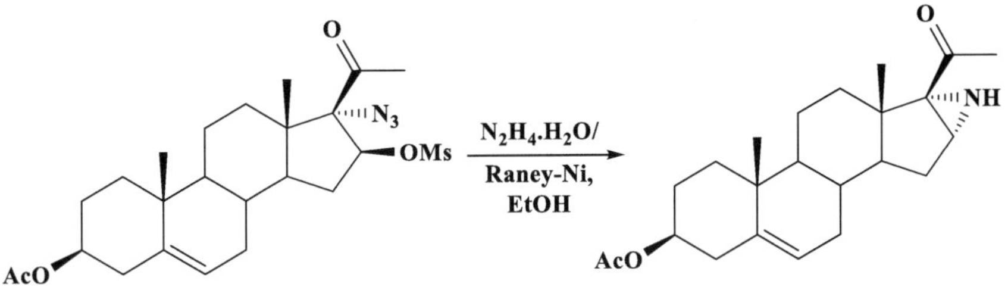

Scheme 3.40

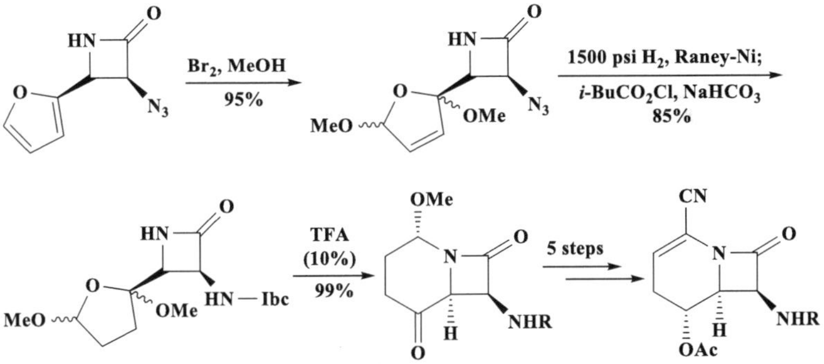

Scheme 3.41

substitution involving the reduction of diastereomeric 16β-mesyloxy-17α-azidopregnenon provided steroidal acylaziridine (Scheme 3.40) [97,98].

Ciufolini and Dong [99] utilized this procedure for the formation of cyanocarbacephem (Scheme 3.41), which is an example of antibiotic class of carbacephems. The dimethylacetal was obtained in 95% yield by reacting β-lactam with bromine in a three-step process. Then, the olefin hydrogenation over Raney nickel led to the simultaneous reduction of azide moiety. The amine was protected as an *iso*-butyl carbamate (Ibc), after which aza-Achmatowicz product was obtained in 99% yield by the hydrolysis of acetals. The cyanocarbacephem was prepared by five additional synthetic reactions. An excellent review was published by Ciufolini et al. [100] on their efforts on the aza-Achmatowicz reaction.

References

[1] (a) Kaur N. Six-membered *N*-heterocycles: microwave-assisted synthesis. Synth Commun 2015;45:1–34; (b) Kaur N. Polycyclic six-membered *N*-heterocycles: microwave-assisted synthesis. Synth Commun 2015;45:35–69; (c) Kaur N. Copper catalysts in

the synthesis of five-membered *N*-polyheterocycles. Curr Org Synth 2018;15:940–71; (d) Kaur N. Recent impact of microwave-assisted synthesis on benzo derivatives of five-membered *N*-heterocycles. Synth Commun 2015;45:539–68; (e) Kaur N. Gold catalysts in the synthesis of five-membered *N*-heterocycles. Curr Organocatal 2017;4: 122–154.
[2] (a) Kaur N. Microwave-assisted synthesis: fused five-membered *N*-heterocycles. Synth Commun 2015;45:789–823; (b) Kaur N. Six-membered heterocycles with three and four *N*-heteroatoms: microwave-assisted synthesis. Synth Commun 2015;45:151–72; (c) Kaur N. Application of microwave-assisted synthesis in the synthesis of fused six-membered heterocycles with *N*-heteroatom. Synth Commun 2015;45:173–201; (d) Kaur N, Kishore D. Microwave-assisted synthesis of six-membered *O,O*-heterocycles. Synth Commun 2014;44:3082–111; (e) Kaur N, Kishore D. Microwave-assisted synthesis of six-membered *O*-heterocycles. Synth Commun 2014;44:3047–81.
[3] (a) Kaur N. Microwave-assisted synthesis of fused polycyclic six-membered *N*-heterocycles. Synth Commun 2015;45:273–99; (b) Kaur N. Review of microwave-assisted synthesis of benzo-fused six-membered *N,N*-heterocycles. Synth Commun 2015;45:300–30; (c) Kaur N, Kishore D. Synthetic strategies applicable in the synthesis of privileged scaffold: 1,4-benzodiazepine. Synth Commun 2014;44:1375–413; (d) Kaur N. Synthesis of five-membered *N,N,N*- and *N,N,N,N*-heterocyclic compounds: applications of microwaves. Synth Commun 2015;45:1711–42; (e) Kaur N. Palladium acetate and phosphine assisted synthesis of five-membered *N*-heterocycles. Synth Commun 2019;49:483–514; (f) Kaur N. Ionic liquid: an efficient and recyclable medium for the synthesis of fused six-membered oxygen heterocycles. Synth Commun 2019;49:1679–707.
[4] (a) Kaur N. Environmentally benign synthesis of five-membered 1,3-*N,N*-heterocycles by microwave irradiation. Synth Commun 2015;45:909–43; (b) Kaur N. Advances in microwave-assisted synthesis for five-membered *N*-heterocycles synthesis. Synth Commun 2015;45:432–57; (c) Kaur N. Multiple nitrogen-containing heterocycles: metal and non-metal assisted synthesis. Synth Commun 2019;49:1633–58; (d) Kaur N, Grewal P, Bhardwaj P, Devi M, Verma Y. Nickel-catalyzed synthesis of five-membered heterocycles. Synth Commun 2019;49:1543–77; (e) Tyagi R, Kaur N, Singh B, Kishore D. A novel synthetic protocol for the heteroannulation of oxocarbazole and oxoazacarbazole derivatives through corresponding oxoketene dithioacetals. J Heterocycl Chem 2014;51:18–23.
[5] Balaban AT, Oniciu DC, Katritzky AR. Aromaticity as a cornerstone of heterocyclic chemistry. Chem Rev 2004;104:2777–812.
[6] Martins MAP, Cunico W, Pereira CMP, Flores AFC, Bonacorso HG, Zanatta N. 4-Alkoxy-1,1,1-trichloro-3-alken-2-ones: preparation and applications in heterocyclic synthesis. Curr Org Synth 2004;1:391–403.
[7] Druzhinin SV, Balenkova ES, Nenajdenko VG. Recent advances in the chemistry of α,β-unsaturated trifluoromethylketones. Tetrahedron 2007;63:7753–808.
[8] Boaen NK, Hillmyer MA. Post-polymerization functionalization of polyolefins. Chem Soc Rev 2005;34:267–75.
[9] Koch MA, Schuffenhauer A, Scheck M, Wetzel S, Casaulta M, Odermatt A, Ertl P and Waldmann H. Charting biologically relevant chemical space: a structural classification of natural products (SCONP). Proc. Natl. Acad. Sci. USA 102: 17272-17277.
[10] Hili R, Yudin AK. Making carbon-nitrogen bonds in biological and chemical synthesis. Nat Chem Biol 2006;2:284–7.
[11] Carey JS, Laffan D, Thomson C, Williams MT. Analysis of the reactions used for the preparation of drug candidate molecules. Org Biomol Chem 2006;4:2337–47.
[12] Tohme R, Darwiche N, Gali-Muhtasib H. A journey under the sea: the quest for marine anti-cancer alkaloids. Molecules 2011;16:9665–96.

[13] von Braun J, Blessing G, Zobel F. Catalytic hydrogenations under pressure in the presence of nickel salts. VI: Nitriles Ber Dtsch Chem Ges 1923;56:1988–2001.
[14] Stetter H, Schwarz M. Synthese des 4.5.9.10-tetraaza-pyrens. Chem Ber 1957;90: 1349–1351.
[15] Nate H, Sekine Y, Oda K, Aoe K, Nakai H, Wada H, Takeda M, Yabana H, Nagao T. Synthesis of 2-phenylthiazolidine derivatives as cardiotonic agents. V. Modification of the thiazolidine moiety of 2-(phenylpiperazinylalkoxyphenyl)thiazolidine-3-thiocarboxamides and the corresponding carboxamides. Chem Pharm Bull 1987;35:3253–61.
[16] Vries T, Wynberg H, van Echten E, Koek J, ten Hoeve W, Kellogg RM, Broxterman QB, Minnaard A, Kaptein B, van der Sluis S, Hulshof L. The family approach to the resolution of racemates. Angew Chem, Int Ed 1998;37:2349–54.
[17] Sperber N, Fricano R. Cyclic derivatives of α,α-disubstituted phenylacetonitriles. J Am Chem Soc 1953;75:2986–8.
[18] Fischer H, Plieninger H. Synthese des biliverdins (uteroverdins) und bilirubins, der biliverdine XIIIα und IIIα, sowie der vinylneoxanthosäure. Hoppe-Seyler's. Physiol Chem 1942;274:231–60.
[19] Fischer H, Plieninger H. Synthese des biliverdins (uteroverdins) und bilirubins. Naturwissenschafren 1942;30:382–7.
[20] Fischer H, Lautsch W. Teilsynthese von methylphäophorbid a und methylphäophorbid b. Liebigs Ann Chem 1937;528:265–75.
[21] Plieninger H, Decker M. Eine neue synthese für pyrrolone, insbesondere für "isooxyopsopyrrol" und "isooxyopsopyrrol-carbonsäure". Liebigs Ann Chem 1956; 598:198–207.
[22] Plieninger H, Kune J. Synthese der "oxyopsopyrrolcarbonsäure" und weitere untersuchungen in der pyrrolon-reihe. Liebigs Ann Chem 1964;680:60–9.
[23] Boiadjiev SE, Lightner DA. Dipyrrinones - constituents of the pigments of life. A review. Org Prep Proced 2006;38:347–99.
[24] Comins DL, Schilling S, Zhang Y. Asymmetric synthesis of 3-substituted isoindolinones: application to the total synthesis of (+)-lennoxamine. Org Lett 2005;7:95–8.
[25] Kametani T, Honda T, Inoue H, Fukumoto KA. One-step synthesis of the phthalidylisoquinoline alkaloids, cordrastine and hydrastine. J Chem Soc, Perkin Trans 1 1976;11:1221–5.
[26] Leonard MS. The *Aporhoeadane* alkaloids (13-7648LR). ARKIVOC 2013(i):1–65.
[27] Peglion J-L, Poitevin C, Cour CML, Dupuis D, Millan MJ. Modulations of the amide function of the preferential dopamine D3 agonist (*R,R*)-S32504: improvements of affinity and selectivity for D3 versus D2 receptors. Bioorg Med Chem Lett 2009;19:2133–8.
[28] Khurana JM, Kukreja G. Rapid reduction of nitriles to primary amines with nickel boride at ambient temperature. Synth Commun 2002;32:1265–9.
[29] Hara O, Sugimoto K, Hamada Y. Synthetic studies on bradykinin antagonist martinellines: construction of a pyrrolo[3,2-c]quinoline skeleton using silicon-tether RCM reaction and allylic amination. Tetrahedron 2004;60:9381–90.
[30] Sridharan V, Suryavanshi PA, Menendez JC. Advances in the chemistry of tetrahydroquinolines. Chem Rev 2011;111:7157–259.
[31] Beemelmanns C, Reissig HU. A short formal total synthesis of strychnine with a samarium diiodide induced cascade reaction as the key step. Angew Chem, Int Ed 2010;49:8021–5.
[32] Ishikura M, Abe T, Choshi T, Hibino S. Simple indole alkaloids and those with a non-rearranged monoterpenoid unit. Nat Prod Rep 2013;30:694–752.
[33] Farcasiu D. Solvolysis of 1-substituted-2-adamantyl sulfonates. Bridging or no bridging? J Am Chem Soc 1976;98:5301–5.

[34] Oldenziel O, van Leusen D, van Leusen A. Chemistry of sulfonylmethyl isocyanides. 13. A general one-step synthesis of nitriles from ketones using tosylmethyl isocyanide. Introduction of a one-carbon unit. J Org Chem 1977;42:3114–18.
[35] Badger GM, Cook JW, Walker T. Synthesis of piperidine derivatives. Part II. Aryldecahydroquinolines. J Chem Soc 1948;0:2011–17.
[36] Kondyukov IZ, Karpychev YV, Belyaev PG, Khisamutdinov GK, Valeshnii SI, Smirnov SP, Il'in VP. Sulfur as a new low-cost and selective reducing agent for the transformation of benzofuroxans into benzofurazans. Russ J Org Chem 1979;43:635–6.
[37] Moiseev IK, Konovalova VP. Adamantane and its derivatives. Hydrazones and pyrazolones of adamantane. Chem Heterocycl Compd (Engl Transl) 1982;18:405–8.
[38] Scheibye S, Kristensen J, Lawesson SO. Studies on organophosphorus compounds - XXVII. Tetrahedron 1979;35:1339–43.
[39] Zoidis G, Tsotinis A, Kolocouris N, Kelly JM, Prathalingam SR, Naesens L, de Clercq E. Design and synthesis of bioactive 1,2-annulated adamantane derivatives. Org Biomol Chem 2008;6:3177–85.
[40] Aurins A, Ankhou V. Isothiazolopyridines. Synthesis and spectra of isothiazolo[3,4-b]-, 3-amino-isothiazolo[4,3-b]-, isothiazolo[5,4-b]-, and 3-methylisothiazolo[5,4-c]pyridines. Preparation and spectra of some 2,3- and 3,4-disubstituted pyridines. Can J Chem 1973;51:1741–8.
[41] Singh V, Yadav GP, Maulik PR, Batra S. Studies toward the construction of substituted piperidine-2-ones and pyridine-2-ones from Baylis-Hillman adducts: discovery of a facile synthesis of 5-methyl-4-oxo-6-aryl-3-aza-bicyclo[3.1.0]hexane-1-carboxylates. Tetrahedron 2006;62:8731–9.
[42] Singh V, Yadav GP, Maulikb PR, Batra S. Synthesis of substituted 3-methylene-2-pyridones from Baylis-Hillman derivatives and its application for the generation of 2-pyridone substituted spiroisoxazolines. Tetrahedron 2008;64:2979–91.
[43] Zhong W, Liu Y, Wang G, Hong L, Chen Y, Chen X, Zheng Y, Zhang W, Ma W, Shen Y, Yao Y. Recent advances in construction of nitrogen-containing heterocycles from Baylis-Hillman adducts. Org Prep Proced Int 2011;43:1–66.
[44] Albertson NF. Piperidines and azabicyclo compounds. I. Via Michael condensations. J Am Chem Soc 1950;72:2594–9.
[45] Badger GM, Cook JW, Walker T. The synthesis of piperidine derivatives. Part III. 5-Phenyl-1-azabicyclo[3.3.1]nonane. Chem Commun 1949;0:1141–4.
[46] Albertson NF. So-called 5-carbethoxy-9-methyl-2-oxo-1-azabicyclo[3.3.1]nonane. A correction. J Am Chem Soc 1952;74:249–51.
[47] Szostak M, Aube J. Chemistry of bridged lactams and related heterocycles. Chem Rev 2013;113:5701–65.
[48] Dwyer MP, Lamar JE, Meyers AI. Synthesis of enantiomerically pure cis-2,4-disubstituted piperidines: extension of chiral homoenolate alkylations toward the preparation of nitrogen heterocycles. Tetrahedron Lett 1999;40:8965–8.
[49] Taylor MS, Zalatan DN, Lerchner AM, Jacobsen EN. Highly enantioselective conjugate additions to α,β-unsaturated ketones catalyzed by a (salen)Al complex. J Am Chem Soc 2005;127:1313–17.
[50] Baylis AB and Hillman MED. 1972. System and method for preparing a corneal graft. German Patent 2155113. CA, 77, 34174q.
[51] Patra A, Batra S, Kundu B, Joshi BS, Roy R, Bhaduri AP. 5-Isoxazolecarboxaldehyde: a novel substrate for fast Baylis-Hillman reaction. Synthesis 2001;2:276–80.
[52] Singh V, Batra S. Convenient synthesis of substituted α-methylene-δ-valerolactones in aqueous medium using Baylis-Hillman chemistry. Synthesis 2006;1:63–72.
[53] Thomas OP, Dumas C, Zaparucha A, Husson HP. Synthesis of ($\pm$)-17-methylcamptothecins. Eur J Org Chem 2004;5:1128–35.

[54] Prugh JD, Deana AA, Wiggins JA. Convenient and unambiguous synthesis of 2 (or 7)-chloronaphthalenes from substituted α-tetralones. Synthesis 1989;7:554–6.
[55] Data Centre, 12 Union Road, Cambridge, CB2 1EZ, UK (CCDC deposition no. of 9e: 609070).
[56] Wang S, Tan T, Li J, Hu H. Highly efficient one-pot synthesis of 1,2-dihydro-2-oxo-3-pyridine-carboxylate derivatives by $FeCl_3$-promoted [3+3] annulation. Synlett 2005;17:2658–60.
[57] Lee MJ, Kim SC, Kim JN. The first synthesis of 3,5-dimethylene-4-phenylpiperidine-2,6-dione from Baylis-Hillman adduct. Bull Korean Chem Soc 2006;27:140–2.
[58] Evans DA, Scheerer JR. Polycyclic molecules from linear precursors: stereoselective synthesis of clavolonine and related complex structures. Angew Chem, Int Ed 2005;44:6038–42.
[59] Vardanyan RS, Hruby VJ. Antiarrhythmic drugs. Synthesis of essential drugs. Elsevier; 2006. p. 245–55.
[60] Cliffe IA, Todd RS, White AC. Synthesis of 2,2-dialkyl-3-haiopropannitriles from 2,2-dialkylethanenitrles and dihalomethanes. Synth Commun 1990;20:1757–67.
[61] Havran L, Chong DC, Childers WE, Dollings PJ, Dietrich A, Harrison BL, Marathias V, Tawa G, Aulabaugh A, Cowling R, Kapoor B, Xu W, Mosyak L, Moy F, Hum W-T, Wood A, Robichaud AJ. 3,4-Dihydropyrimido(1,2-a)indol-10(2H)-ones as potent non-peptidic inhibitors of caspase-3. Bioorg Med Chem 2009;17:7755–68.
[62] Udemba A. PhD Thesis. Imperial College London; 2013. p. 1–247.
[63] Kumar KP, Rao JV, Mukkanti K, Raju MB, Khan KA. One-pot synthesis characterization and evaluation of some novel pyrazolopyrimidines derivatives using 1-ethyl-3-(3-dimethylaminopropyl)carbodiimide (EDCI), as coupling agent as potential anticatotonic agents. J Pharm Res 2010;3:1021–4.
[64] Lenda F, Guenoun F, Tazi B, Larbi NB, Allouchi H, Martinez J, Lamaty F. Synthesis of new tetrazole-substituted pyroaminoadipic and pipecolic acid derivatives. Eur J Org Chem 2005;2:326–33.
[65] McKay AF, Braun RO. Cyclizations of β-chloroethyl substituted ammonocarbonic acids. J Org Chem 1951;16:1829–34.
[66] Depreux P, Lesieur D, Mansour HA, Morgan P, Howell HE, Renard P, Caignard DH, Pfeiffer B, Delagrange P, Guardiola B. Synthesis and structure-activity relationships of novel naphthalenic and bioisosteric related amidic derivatives as melatonin receptor ligands. J Med Chem 1994;37:3231–9.
[67] Mésangeau C, Yous S, Pérèz B, Lesieur D, Besson T. Pictet-Spengler heterocyclizations via microwave-assisted degradation of DMSO. Tetrahedron Lett 2005;46: 2465–2468.
[68] Rami M, Landagaray E, Ettaoussi M, Boukhalfa K, Caignard D-H, Delagrange P, Berthelot P, Yous S. Novel conformationally constrained analogues of agomelatine as new melatoninergic ligands. Molecules 2013;18:154–66.
[69] Stankevichyus AP, Kost AN. Reductive cyclization of o-cyanocinnamic acids and their analogs. Chem Heterocycl Compd 1971;7:1208–12.
[70] Zhukauskaite LN, Stankevichus AP, Kost AN. Synthesis of condensed structures from 3-cyano-2-pyridylacrylic acids. Chem Heterocycl Compd 1978;14:49–54.
[71] (a) Wenner W. 6,7-Dihydro-5H-dibenz[c,e]azepine derivatives, a new class of epinephrine antagonists. J Org Chem 1951;16:1475–80; (b) Jeffs PN, Hansen JF. Synthesis of a medium ring containing bridge biphenyl by photochemically induced intramolecular arylation. J Am Chem Soc 1967;89:2798–9.
[72] Ishida Y, Sasaki Y, Kimura Y, Watanabe K. Alpha-adrenolytic properties of apogalanthamine and azapetine analogs. J Pharmacobiodyn 1985;8:917–23.
[73] Stankevicius AP, Yanushene LN. Synthesis of 6,7-dihydro-5H-dibenz[c,e]azepine. Chem Heterocycl Compd 2006;42:1488–9.

[74] Herrero S, García-López MT, Herranz H. Expedient one-pot synthesis of novel chiral 2-substituted 5-phenyl-1,4-benzodiazepine scaffolds from amino acid-derived amino nitriles. J Org Chem 2003;68:4582–5.
[75] Singh P, Samanta K, Das SK, Panda G. Amino acid chirons: a tool for asymmetric synthesis of heterocycles. Org Biomol Chem 2014;12:6297–339.
[76] van Berkom LWA. PhD Thesis. The Netherlands: Radboud Universiteit Nijmegen; 2005.
[77] Kamal A, Ramulu P, Srinivas O, Ramesh G. Synthesis and DNA-binding affinity of A-C8/C-C2 alkoxyamido-linked pyrrolo[2,1-c][1,4]benzodiazepine dimers. Bioorg Med Chem Lett 2003;13:3955–8.
[78] Capon RJ, Skene C, Ford J, O'Hair RAJ, Williams L, Lacey E, Gill JH, Heiland K, Friedel T. Aspergillicins A-E: five novel depsipeptides from the marine-derived fungus *Aspergillus carneus*. Org Biomol Chem 2003;1:1856–62.
[79] Kamal A, Ramesh G, Laxman N, Ramulu P, Srinivas O, Neelina K, Kondapi AK, Sreenu VB, Nagarajaram HA. Design, synthesis, and evaluation of new noncrosslinking pyrrolobenzodiazepine dimers with efficient DNA binding ability and potent antitumor activity. J Med Chem 2002;45:4679–88.
[80] Humphrey AJ, Parsons SF, Smith MEB, Turner NJ. Synthesis of a novel N-hydroxypyrrolidine using enzyme catalysed asymmetric carbon-carbon bond synthesis. Tetrahedron Lett 2000;41:4481–5.
[81] Kayakiri H, Takase S, Seloi H, Uchida I, Terano H, Hashimoto M. Structure of FR 900483, a new immunomodulator isolated from a fungus. Tetrahedron Lett 1988;29:1725–8.
[82] Kayakiri H, Nakamura K, Takase S, Seloi H, Uchida I, Terano H, Hashimoto M, Tada T, Koda S. Structure and synthesis of nectrisine, a new immunomodulator isolated from a fungus. Chem Pharm Bull 1991;39:2807–12.
[83] Chakrabarti J, Szinai S, Todd A. Chemistry of adamantane. Part III. The synthesis and reactions of 1,2-disubstituted adamantane derivatives. J Chem Soc C 1970;9:1303–9.
[84] Chakrabarti J, Hotten T, Rackham D, Tupper D. Chemistry of adamantane. Part IX. 1,2-Difunctional adamantanes: synthesis and reactions of protoadamantane-4-spirooxiran. J Chem Soc, Perkin Trans 1 1976;17:1893–900.
[85] Lunn WHW, Podmore WD, Szinai S. Adamantane chemistry. Part I. The synthesis of 1,2-disubstituted adamantanes. J Chem Soc C 1968;0:1657–60.
[86] Kolocouris N, Zoidis G, Fytas C. Facile synthetic routes to 2-oxo-1-adamantanalkanoic acids. Synlett 2007;7:1063–6.
[87] Clauson-Kaas N, Nedenskov P. Preparation of certain methyl polyhydroxybenzoates from methyl furoate. Acta Chem Scand 1955;9:27–9.
[88] Williams H, Kaufmann P, Mosher HS. The rearrangement of acylfurans to 3-hydroxypyridines. J Org Chem 1955;20:1139–45.
[89] Kaniskan HU. PhD Thesis. Case Western Reserve University; 2007.
[90] Ishikawa F, Kosayama A and Abiko K. 1978. JP Patent 53 044 592.
[91] Ishikawa F, Kosayama A and Abiko K. 1978. JP Patent 53 044 593.
[92] Dumitrascu F, Popa MM. Pyrrolo[1,2-a]quinazolines: synthesis and biological properties (14-8699LR). ARKIVOC 2014(i):428–52.
[93] Baddiley J, Lythgoe B, McNeil D, Todd AR. Experiments on the synthesis of purine nucleosides. Part I. Model experiments on the synthesis of 9-alkylpurines. J Chem Soc 1943;0:383–6.
[94] Reisman SE, Ready JM, Hasuoka A, Smith CJ, Wood JL. Total synthesis of (±)-welwitindolinone A isonitrile. J Am Chem Soc 2006;128:1448–9.
[95] Reisman SE, Ready JM, Weiss MM, Hasuoka A, Hirata M, Tamaki K, Ovaska TV, Smith CJ, Wood JL. Evolution of a synthetic strategy: total synthesis of (±)-welwitindolinone A isonitrile. J Am Chem Soc 2008;130:2087–100.

[96] Ackrell J, Muchowski JM, Galeazzi E, Guzman A. Alkylation of alpha-formamido ketone enolate anions. A versatile synthesis of alpha-alkyl alpha-amino ketones. J Org Chem 1986;51:3374–6.
[97] Ponsold K, Schoenecker B, Pfaff I. Nitrogen containing steroids, XVII. 17α-Azido and 17α-amino pregnanes. Chem Ber 1967;100:2957–66.
[98] Patonay T, Konya K, Juhasz-Toth E. Synthesis and transformations of α-azido ketones and related derivatives. Chem Soc Rev 2011;40:2797–847.
[99] Ciufolini MA, Dong Q. Aza-Achmatowicz route to novel cyanocarbacephems. Chem Commun 1996;7:881–2.
[100] Ciufolini MA, Hermann CYW, Dong Q, Shimizu T, Swaminathan S, Xi N. Nitrogen heterocycles from furans: the aza-Achmatowicz reaction. Synlett 1998;2:105–14.

CHAPTER 4

Synthesis of heterocycles from oxazoles and oxazines using Raney nickel

4.1 Introduction

Heterocycles are ubiquitous in organic materials, pharmaceuticals, various functional molecules, and natural products. As a result, the ongoing interest for developing new efficient and versatile preparation of heterocycles has always been a thread in the synthetic area. In the last years the creative ideas of multicomponent procedures, domino reactions and sequential reactions, where complex and highly diverse structures are synthesized in a one-pot manner, have significantly stimulated both industry and academia [1–14].

Although the synthesis of heterogeneous catalysts is usually more direct than their homogeneous equivalents, a degree of particular information is mandatory for the preparation of solid catalysts. Heterogeneous catalysts possess many properties, like the metal crystallite size (distribution), the pore size of support, the specific surface area of metal and its support, the particle size of promoter, the catalyst, and also impurities. Special precautions must be taken in catalyst synthesis because each of these factors influences the catalytic selectivity and activity. For instance, various raw materials and synthetic procedures yield diverse catalysts, even if labeled by the same name (e.g., 20% nickel/aluminum oxide = 20% Ni loaded on Al_2O_3). Therefore, correct catalyst synthesis approaches and raw compounds must be selected for any intended reaction. The ease with which the catalyst is isolated, recycled, and scaled-up for the reaction are generally claimed to be the benefits of heterogeneous catalysts, which can be utilized in flow reactors that consecutively synthesize the products or, in the case of the gas-phase reaction or fast reaction, in the liquid-phase. However, the heterogeneous catalysts have some drawbacks. The reactions performed over heterogeneous catalysts occasionally need specialized apparatus (not generally for laboratory-based organic synthesis) - maybe a stainless steel autoclave for high pressure reactions or/and a specialized equipment for the gas feed. Likewise, an

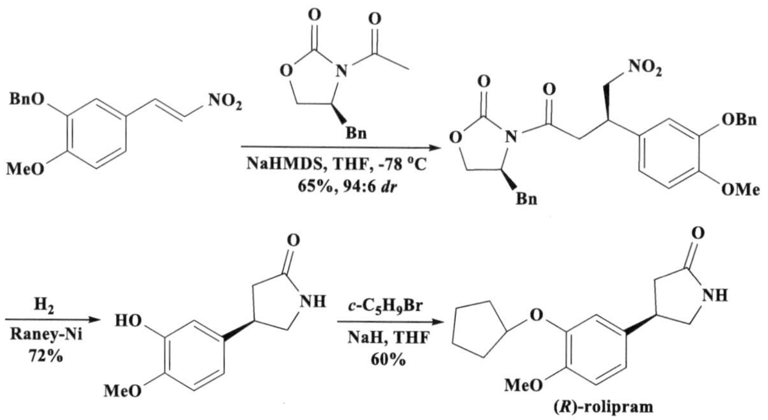

Scheme 4.1

investigation of the reaction mechanism and the formation of single type of active sites on solid are more difficult to carry out for heterogeneous reactions in comparison to homogeneous. Murray Raney discovered Raney-Ni in 1927, and is the broadly utilized unsupported nickel catalyst for liquid-phase reactions. It acts as a versatile hydrogenation catalyst in the formation of fine chemicals [15]. Raney-Ni is a good catalyst for the cleavage of N-O bonds in nitroso acetals to afford the 2-hydroxylactams and pyrrolidines [16–18].

4.2 Synthesis of five-membered *N*-heterocycles from oxazoles

The optically active 1,3-oxazolidin-2-ones are known chiral auxiliaries in an asymmetric synthesis. Their *N*-acyl metal enolates were reacted with several functionalized nitroalkenes in highly diastereofacial selective fashion [19]. A practical use of this approach is explained for the enantioselective formation of antidepressant drug (*R*)-rolipram (Scheme 4.1) [20]. The reaction of nitrostyrene derivative and sodium enolate of (*S*)-4-benzyl-3-acetyloxazolidin-2-one afforded adduct with good diastereoselectivity, and this compound was further purified by crystallization. The catalytic hydrogenation ensured the debenzylation of phenolic oxygen and the reduction of NO_2 group to provide the lactam with the release of chiral auxiliary. The formation of (*R*)-rolipram was completed by *O*-alkylation of free OH group. An employment of other systems like (+)-camphor

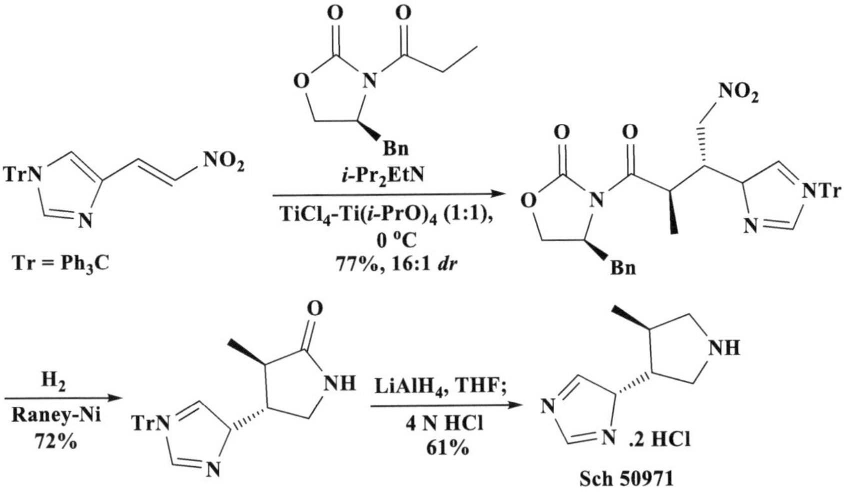

Scheme 4.2

methyl ketone enolates provided results analogous to those observed with 2-oxazolidone on treatment with nitroalkenes, although the cleavage of chiral auxiliary needed oxidative conditions [21].

All the synthetic processes explained for the formation of pyrrolidinones can be utilized to provide the pyrrolidines by the reduction of lactam functionality [22]. The diastereoselective addition of a chiral *N*-propanoyloxazolidin-2-one titanium enolate to imidazolyl nitroolefin was the key step for the formation of Sch 50971, an agonist of histamine H3 receptor (Scheme 4.2) [23]. Further, the lactam group was reduced with lithium aluminum hydride to synthesize the Sch 50971 after the reduction and cyclization to pyrrolidinone.

The formation of Fmoc-protected *trans*-4-methylproline from D-serine derived Garner's aldehyde was described by Nevalainen et al. [24]. The D-serine derived Garner's aldehyde was utilized to control the diastereoselective creation of novel stereocenter by the hydrogenation of allylic alcohol. Use of Raney-Ni afforded 14:86 diastereoselectivity (*anti/syn* ratio). A series of reactions afforded a ring precursor, which was recrystallized to provide the *syn*-diastereoisomer with 5:95 diastereomeric ratio. The protected *trans*-4-methylproline was prepared directly (Scheme 4.3) [25].

Qiu and Qing [25] described a completely different method for the synthesis of 4-Tfm-(*R*)-Pro (Scheme 4.4). The CF$_3$ group was installed first, and after that the proline ring was synthesized. First, the trifluoromethylated

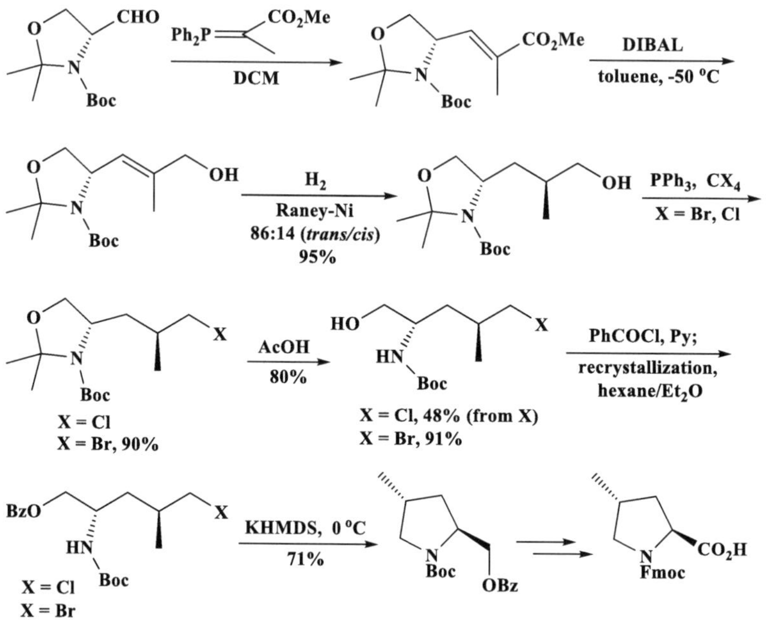

Scheme 4.3

acroleate derivative was prepared using Garner's aldehyde in two steps (not shown). The double bond was reduced by hydrogenolysis (94% yield) and the protection groups were changed in three steps (64% overall yield), and the pure (2R,4S)- and (2R,4R)-isomers were obtained in 3:4 ratio when diastereomers were separated chromatographically. An energetically promising 5-*exo-tet* intramolecular alkylation (80% and 83% yield) resulted in the formation of proline ring after deprotection (91% and 99% yield, respectively), the silyl protecting group was removed quantitatively. The (2R,4S)- and (2R,4R)-isomers of Boc-4-TfmPro were obtained when hydroxymethyl group was oxidized with Jones reagent (56% and 68% yield) [26].

Batra et al. [26,27] converted the Baylis–Hillman adducts derived from 5-isoxazole carboxaldehydes into substituted pyrroles derivatives. They have also converted Baylis-Hillman adducts derived from 3-isooxazole carboxaldehyde to pyrrolidine derivatives (Scheme 4.5) [6].

This sequence was examined using easily accessible model lactone where macrocyclic ring was absent (Scheme 4.6). An addition of C_2H_3MgBr to lactone provided hydroxy ketone in 85% yield [28], which was further

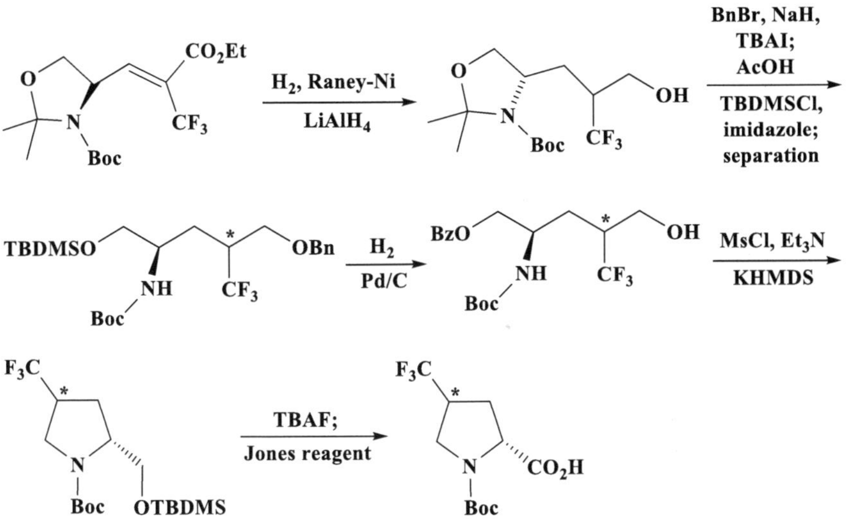

Scheme 4.4

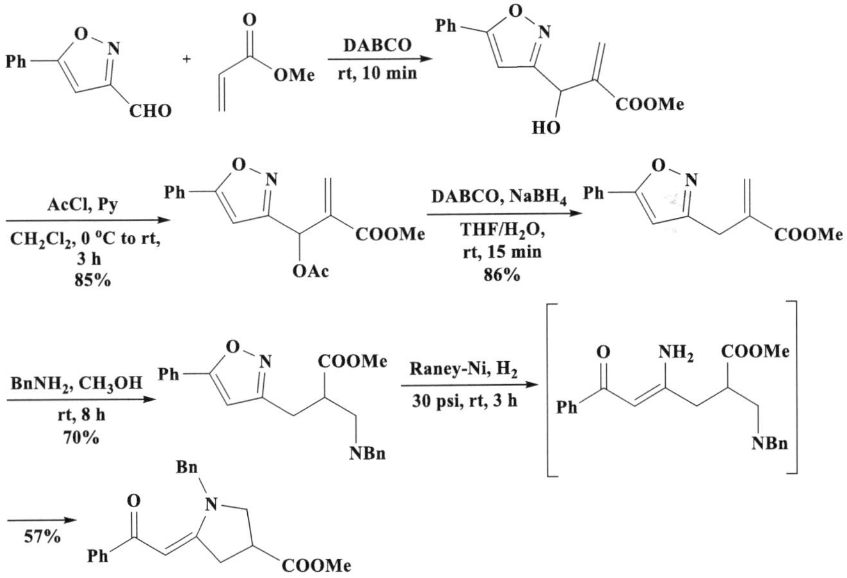

Scheme 4.5

protected as its triethylsilyl ether to provide the enone in 96% yield. The pyrrole ring was comparatively unstable [29] and vulnerable to methylation, therefore, the reaction of enone was examined with benzonitrile-N-oxide. The enone, benzaldehyde oxime, N-chlorosuccinimide, and triethylamine

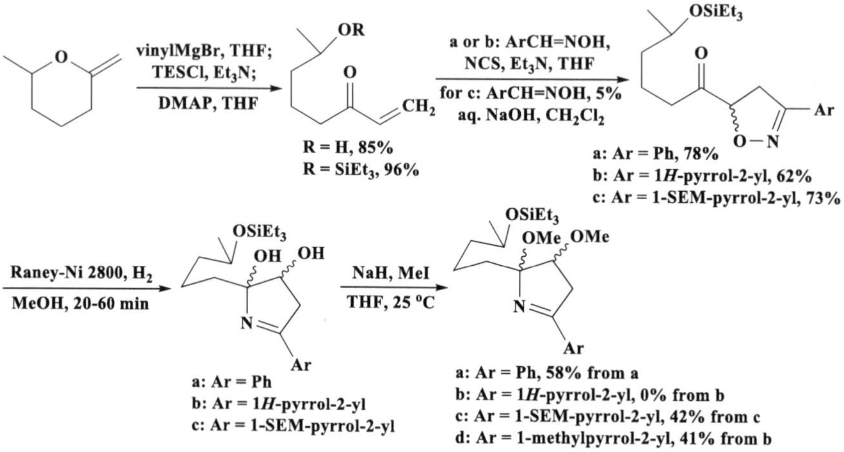

Scheme 4.6

were reacted in tetrahydrofuran at −78 to 25°C to deliver the isoxazoline in 78% yield. The hemi-iminal was obtained as a mixture of isomers by hydrogenolysis (1 atm) over Raney-Ni in methanol for 45 minutes. The methylation with sodium hydride and methyl iodide in tetrahydrofuran at 25°C provided methyl ether iminal in 58% yield from isoxazoline [30].

4.3 Synthesis of five-membered *N*-heterocycles from oxazines

Raney-Ni appropriate for this reaction was prepared from Ni-Al-alloy. The *syn*-oxazine was transformed into *syn*-hydroxy derivative in excellent yield and *anti*-oxazine provided *anti*-hydroxy derivative in 82% yield. These compounds were cyclized to 3-methoxy-2,5-dihydropyrrole derivatives (Scheme 4.7). The reduction of *syn*-hydroxy derivative with hydrogen/palladium-carbon provided a 33:67 mixture of dioxolanes and a 12:88 mixture of dioxolanes was obtained from *anti*-hydroxy derivative (Scheme 4.8). These control experimentations verify that the stereodivergent nature was caused by inversion of the order of reduction steps. The enantiopure dihydropyrroles and pyrrolidine derivatives are important compounds since oxygenated pyrrolidines are known to be biologically active (glycosidase inhibitors etc.) [31–35]. All these compounds originate from two starting compounds, viz. methoxyallene and nitrone, which

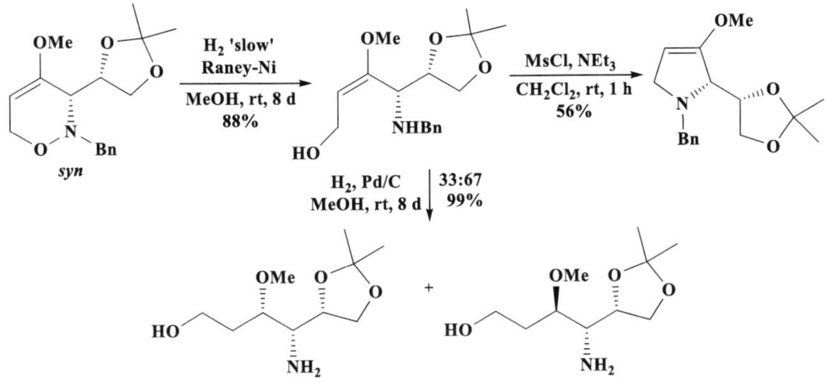

Scheme 4.7

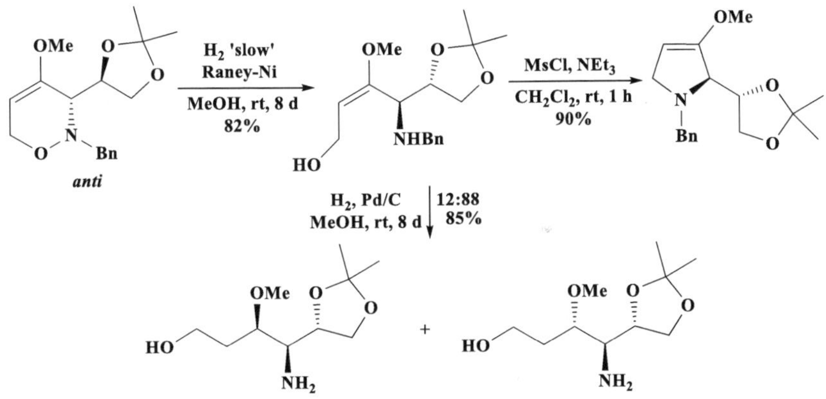

Scheme 4.8

again exhibited the versatility of this C-3 building block for stereoselective formation of heterocyclic compounds [7, 36–39].

The *anti*-oxazine upon hydrogenation using Pd as a catalyst provided a 90:10 mixture of amino alcohols while this selectivity was completely inverted with Raney-Ni (Scheme 4.9). These crude product mixtures were cyclized with MsCl/Et$_3$N to provide the pyrrolidine derivatives in 91:9 ratio. Then, column chromatography was performed for the separation [7].

The N,O-heterocycles are prone to hydrogenolysis [40–43] since the comparatively weak nitrogen–oxygen bond is smoothly cleaved. A solution of *syn*-oxazine in MeOH was reacted with H$_2$ and Pd/C. The crude product had two diastereomeric amino alcohols in 85:15 ratio whose

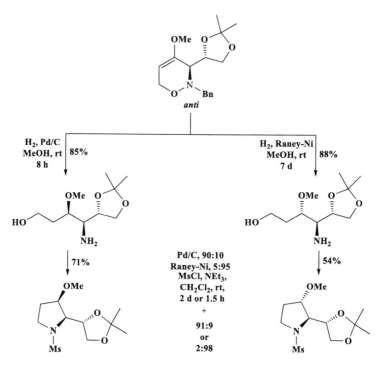

Scheme 4.9

configurations were proved after cyclization. The mixture was reacted with methane sulfonyl chloride using Et$_3$N to provide the 3-methoxypyrrolidines (88:12 ratio). The nuclear Overhauser effect experiments showed the *cis* location of 2-H and 3-H in the major isomer. Remarkably, other compound was the major diastereomer when hydrogenolysis was carried out with Raney-Ni. This slow reaction of *syn*-oxazine afforded a 73:27 ratio, which was converted to cyclized products isolated as a 72:28 mixture (Scheme 4.10). Then, column chromatography was performed for the smooth separation of two isomers [7].

Encouraged by the procedure reported by Denmark and Thorarensen [43], and Kuster [44] used nitro alkenes for the formation of 3-arylpyrrolidines by a [4+2]/[3+2]-cycloaddition reaction (Scheme 4.11).

The C2-substituent in the nitroso acetal was decisive for the synthesis of 2-hydroxy pyrrolizidinone or pyrrolidine framework (Schemes 4.12 – 4.14). Denmark and Thorarensen [16] reported transformations which depend on the solvent and pressure. In this case, the pyrrolidine derivatives were obtained from 2-phenyl-substituted nitroso acetals by a Raney-Ni

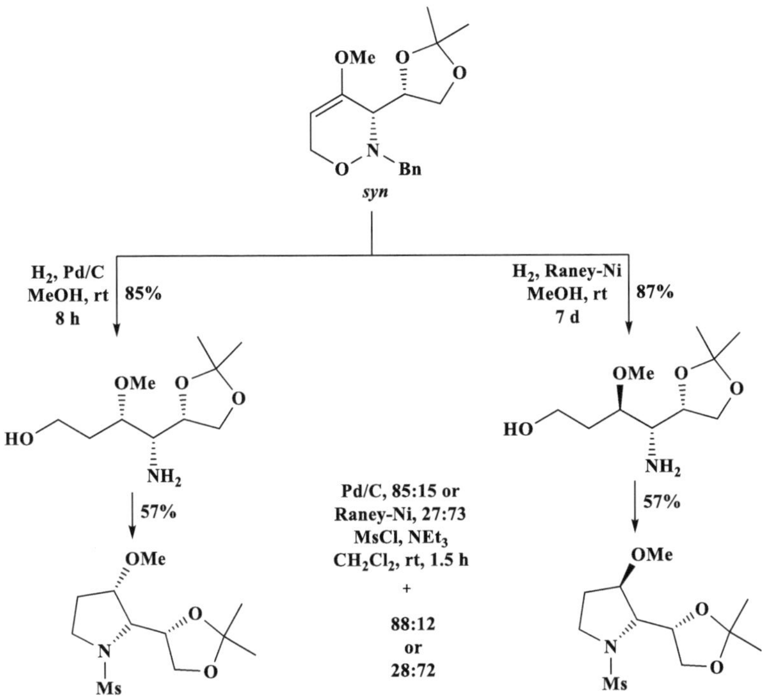

Scheme 4.10

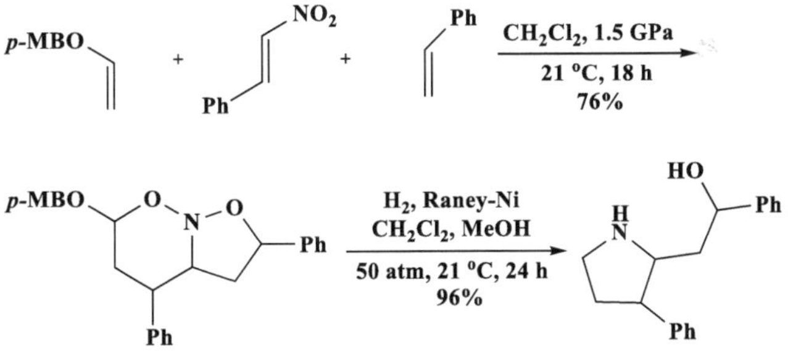

Scheme 4.11

reduction. First, both N-O bonds were hydrogenated and the hemi-acetal intermediate was hydrolyzed to afford the aldehyde. Then, the imine was obtained under Raney-Ni conditions and the pyrrolidine was prepared by the reduction of imine. The pyrrolizidinone derivatives were synthesized from 2-methyl ester-substituted nitroso acetals. The pyrrolidine, obtained by N-O bond reduction, afforded 2-hydroxypyrrolizidinone derivative in a subsequent intramolecular lactam synthesis by elimination of alcohol [17-18].

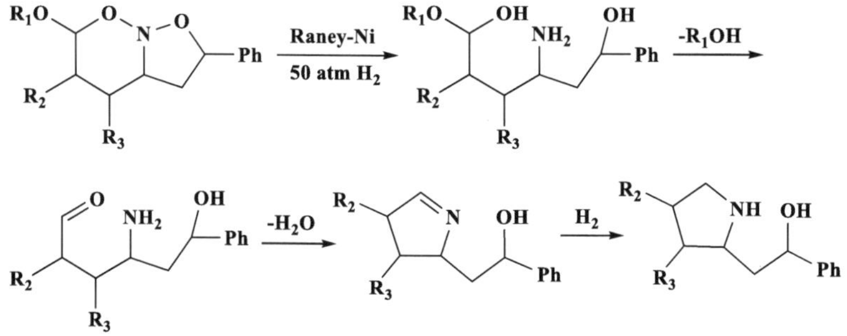

128   Raney nickel-assisted synthesis of heterocycles

$R_1 = $ p-MeOPh, CH_2CH_2, $CH_2CH_2CH_2$
$R_2 = $ H, CH_3
$R_3 = $ Ph, pyrrolyl, pyridyl, indolyl, dihydroxyphenyl

Scheme 4.12

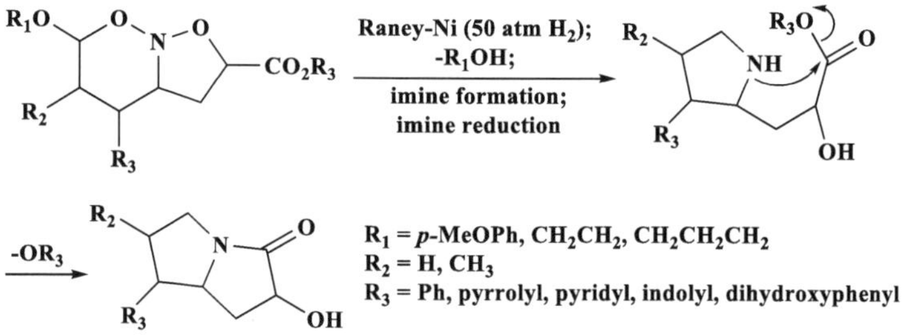

$R_1 = $ p-MeOPh, CH_2CH_2, $CH_2CH_2CH_2$
$R_2 = $ H, CH_3
$R_3 = $ Ph, pyrrolyl, pyridyl, indolyl, dihydroxyphenyl

Scheme 4.13

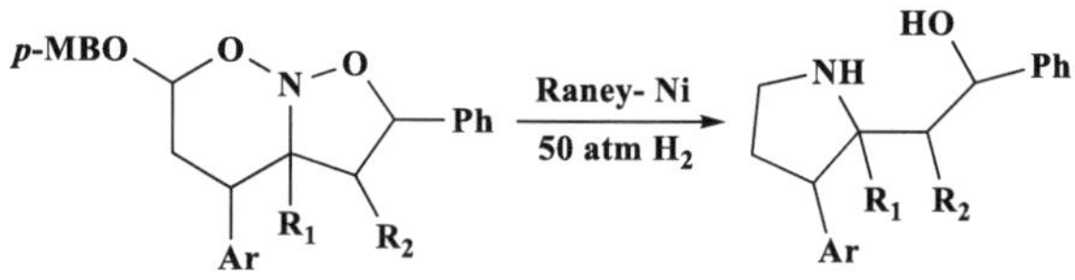

entry	Ar	R_1	R_2	c.y. (%)
1	Ph	H	H	96
2	pyridyl	H	H	44
3	Ph	CH_3	H	95
4	pyrrolyl	CH_3	H	47
5	Ph	CH_3	CO_2Me	95

Scheme 4.14

4.4 Synthesis of five-membered fused N-heterocycles from oxazoles and oxazines

The 1-deoxy-6-*epi*-castanospermine (Scheme 4.15) and 1-deoxy-6,8a-di-*epi*-castanospermine (Scheme 4.16) were obtained when allylic alcohols

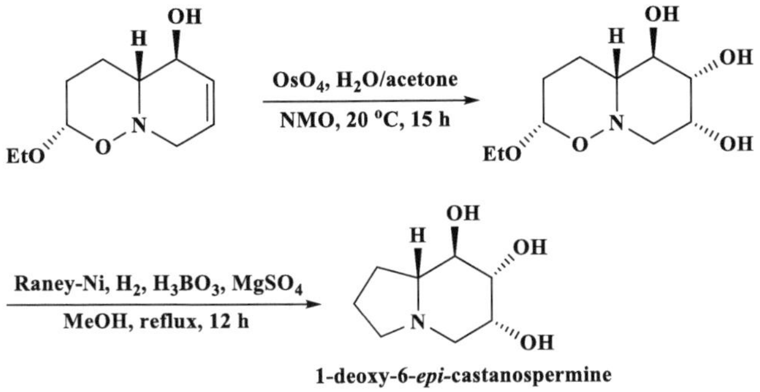

Scheme 4.15

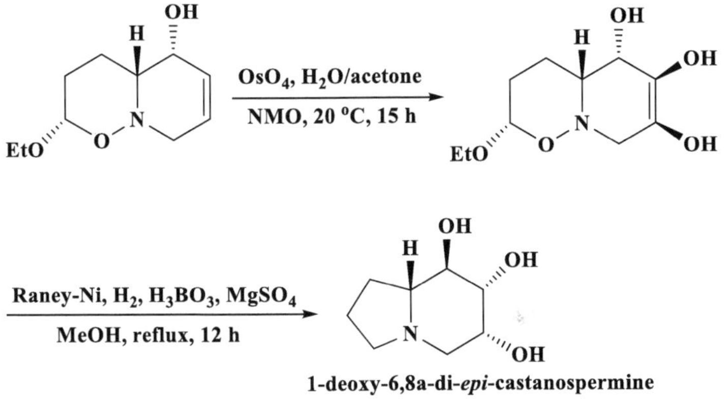

Scheme 4.16

were dihydroxylated by osmylation and then subjected to nitrogen–oxygen bond cleavage with Raney nickel [45].

The mesylate was hydrogenated with Raney-Ni under 160 psi hydrogen pressure in MeOH at rt. Disappointingly, a complex mixture of products was formed, from which only chiral auxiliary was isolated and identified. This result was independent of hydrogen pressure or reaction time; however, the yield of auxiliary was moderate to high. This has proposed that the nitrogen–oxygen bonds cleaved during the reaction, but the cyclization not proceeded as envisioned. The failure to prepare the pyrrolizidine-derived products was because of many complications. First, the leaving group was not suitable or enough active to effect the pyrrolizidine ring-closure. Second, the ring-closure failed because of an inability to access a conformation suitable for the displacement of mesylate by nitrogen, because the two atom tether was present. The cleavage of nitrogen–oxygen bonds revealed a β-OH silicon unit. This offered the

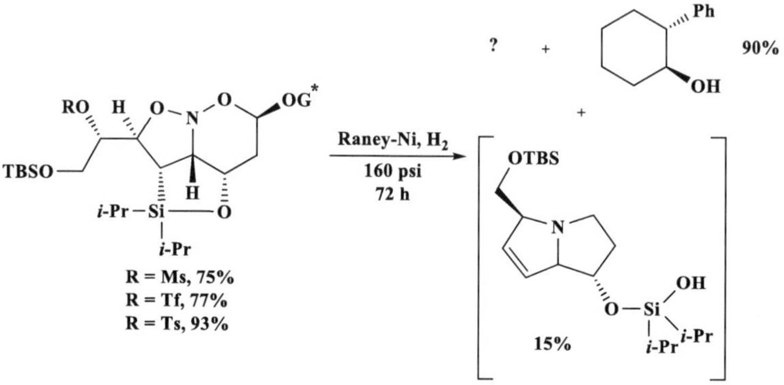

Scheme 4.17

opportunity for a Peterson-type olefination [46] that would cleave the silyl tether, leaving an allylic mesylate (Scheme 4.17). To address the possible difficulties associated to nucleofuge during the ring-closure, many other leaving groups were examined. The tosylate and triflate were prepared from their anhydrides in pyridine. Interestingly, the tosylate solidified after column chromatography. The hydrogenation of triflate in the presence of Raney-Ni occurred with same results as found with mesylate, where no pyrrolizidine bearing products were detected, but the chiral auxiliary was recovered in good yield. However, different behavior was observed in the hydrogenation of tosylate. The recovered chiral auxiliary along with another product was obtained in small amounts. The proton nuclear magnetic resonance spectrum of this product displayed different olefinic peaks, and the structure was tentatively assigned as pyrrolizidine. This has suggested that the Peterson-type olefination procedure interfered with the preferred cyclization route [8].

The 2-hydroxypyrrolizidinones were obtained in moderate to excellent yields by the reduction of 2-methylester-substituted nitroso acetals. The bicyclic 2-hydroxypyrrolizidinones were prepared in good to high yields from dihydroxyphenyl- and phenyl-substituted nitroso acetals (entry 1 and 5), while the 2-hydroxypyrrolizidinones were obtained in lower yields from heteroaryl-substituted nitroso acetals (entry 2, 3, and 4). The pyridyl-substituted 2-hydroxypyrrolizidinone (entry 3) was obtained in the lowest yield (38%), although complete disappearance of the starting material was reported. Circa 80% of *p*-methoxybenzyl alcohol was recovered after chromatographical purification along with some unknown intermediates. Additionally, the indolyl- and pyrrolyl-substituted 2-hydroxypyrro-

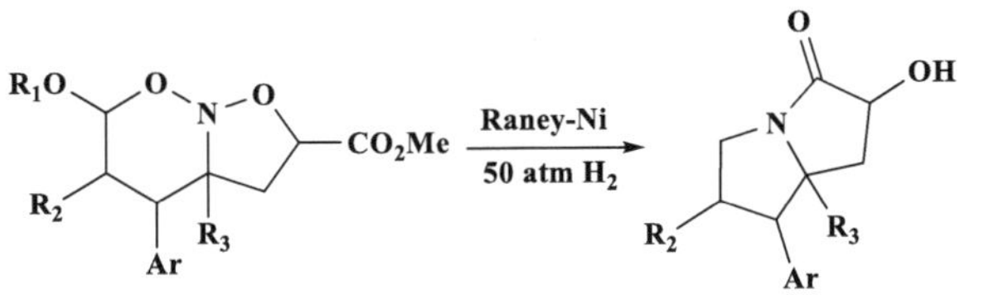

entry	R₁	R₂	Ar	R₃	yield (%)
1	p-MeOPh	H	Ph	H	95
2	p-MeOPh	H	pyrrolyl	H	64
3	p-MeOPh	H	pyridyl	H	38
4	p-MeOPh	H	indolyl	H	52
5	p-MeOPh	H	dihydroxyphenyl	H	74
6	p-MeOPh	CH₃	Ph	CH₃	78
7	CH₂CH₂		Ph	H	88
8	CH₂CH₂CH₂		Ph	H	64
9	CH₂CH₂		Ph	CH₃	71

Scheme 4.18

lizidinones and pyrrolidines were obtained in lower yields than the phenyl-substituted analogues. Perhaps, the elementary nitrogen had an influence on one or more reactions in the reduction cascade. The cycloaddition with dihydrofuran (entry 7 and 9) and dihydropyran (entry 8) resulted in the formation of tricyclic nitroso acetals, which were reduced with Raney-Ni to generate the 6-alkoxy-substituted 2-hydroxypyrrolizidinones in good yields (Scheme 4.18) [44].

Hydrogenolysis of nitroso acetals was difficult to reproduce. The time needed for the complete consumption of starting compound varied considerably and it took over 48 h even under high pressure of hydrogen. In the meantime, the crude products were far from clean and possess several unidentifiable constituents. It was assumed that the unproductive decomposition of starting compounds was caused by the lability of a mesylate in an alcoholic solvent. This difficulty became more serious when the quality of Raney-Ni catalyst was changed. To avoid these complications, it was later observed that a constantly very reactive Raney-Ni was prepared by sonicating the commercial catalyst in deionized H_2O and renewing the H_2O, until the aqueous layer became fully clear and neutral. A mixture of THF and H_2O appeared to be a suitable selection of solvent to allow for smooth hydrogenolysis and decrease the solvolysis of mesylate. Nonetheless, a high pressure of H_2 (400 psi) and a comparatively long reaction time (24 hours) appeared mandatory. These operations afforded two diastereomeric pyrrolizidines (Scheme 4.19) [47].

Denmark and Herbert [48] reported an asymmetric synthesis of 7-*epi*-australine involving an intermolecular nitronate-alkene cycloaddition

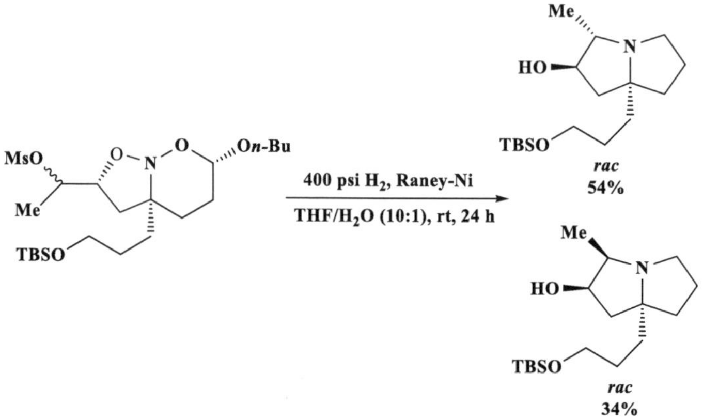

Scheme 4.19

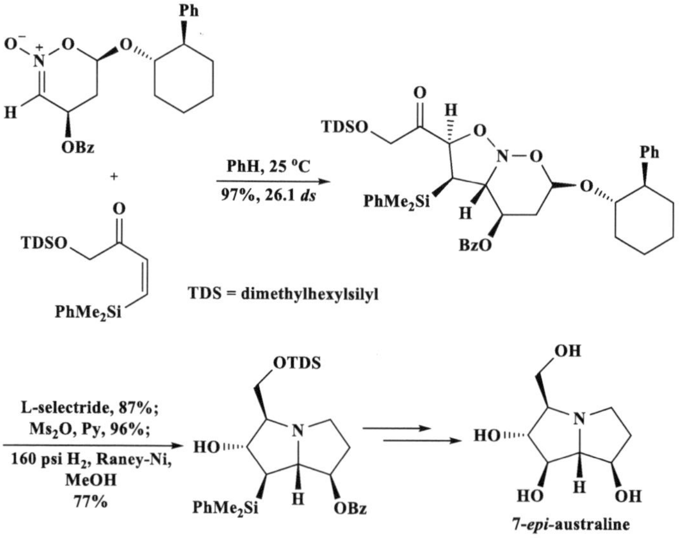

Scheme 4.20

as a key step (Scheme 4.20). The nitronate [synthesized from (1S,2R)-2-phenylcyclohexanol] was treated with vinylsilane to provide the nitroso acetal in 97% yield, almost as a single diastereoisomer. The reduction of ketone with L-selectride, followed by mesylation and hydrogenation led to the cleavage of nitrogen–oxygen bond, cyclization, and reduction to provide the pyrrolizidine in 64% yield. The dimethylphenylsilyl group was then unmasked to introduce the C-1 hydroxy of 7-*epi*-australine. A crystal structure of final product was obtained, which offered explanation of some

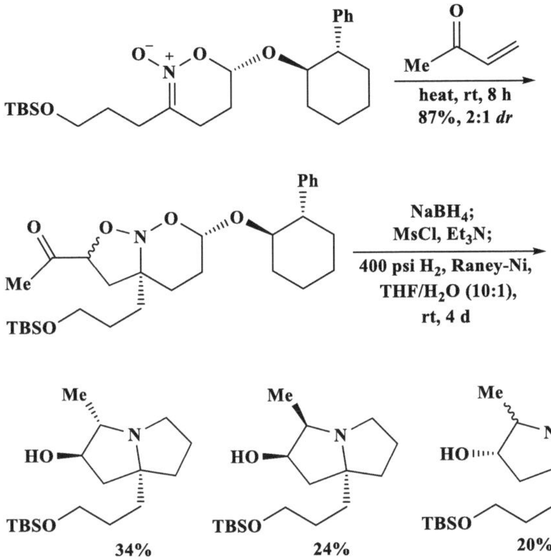

Scheme 4.21

previous misperception in the literature about the exact stereochemistry of 7-*epi*-australine and australine [9].

The nitronate participated in [3+2]-cycloaddition with methyl vinyl ketone. Again, a mixture of diastereomeric nitroso acetals was formed. Disappointingly, the resolution of these two diastereomers became very challenging. Therefore, the mixture was carried on further. Four diastereomers were obtained after the NaBH$_4$ reduction of ketone. Four diastereomeric pyrrolizidines were synthesized by mesylation of alcohols followed by hydrogenolysis. Luckily, each of the two major products could be isolated as a single diastereomer (Scheme 4.21) [47].

The glycosidase inhibitory activity showed by many alkaloids having pyrrolizidine units has encouraged a search for efficient synthetic processes for their formation [49]. The reactivity of nitroalkenes as heterodienes in inverse-electron-demand Diels-Alder reactions is beneficially utilized for the formation of pyrrolidine ring but can be also modified to the formation of more complex *N*-polycyclic derivatives. The [4+2]-cycloadditions of chiral enol ethers and nitroalkenes afforded cyclic nitronates, which on reductive cleavage provided pyrrolidine systems. However, these cyclic nitronates are reactive 1,3-dipoles for [3+2]-cycloadditions with many alkenes like dimethyl maleate for the synthesis of bicyclo derivatives (Scheme 4.22) [16,50].

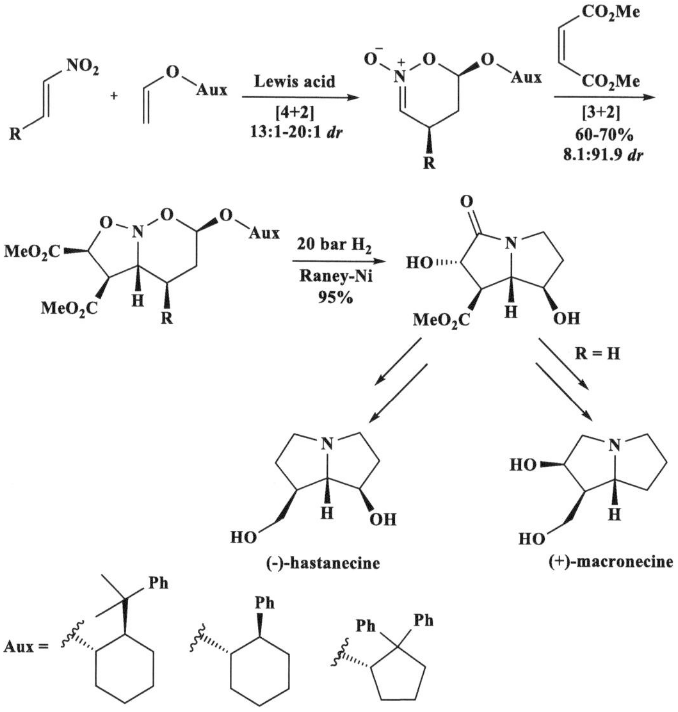

Scheme 4.22

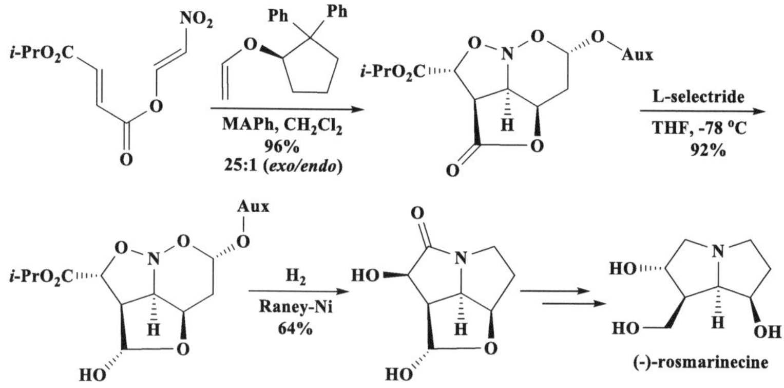

Scheme 4.23

The fumarate-derived nitroalkenes can also participate in tandem intermolecular [4+2]/intramolecular [3+2]-cycloadditions to provide the tricyclic intermediates with satisfactory *exo*-selectivity (Scheme 4.23) [51–52]. For the subsequent reductive rearrangement to pyrrolizidine nucleus, the presence of lactone carbonyl in cycloadduct was problematic

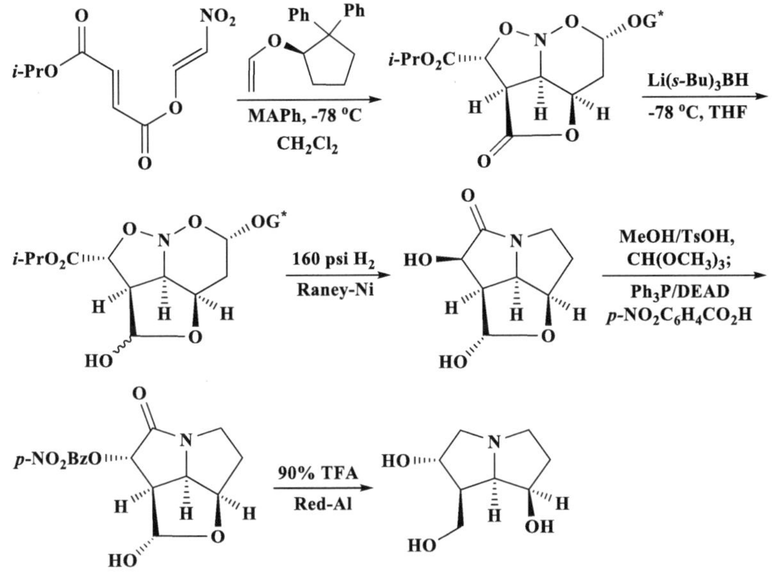

Scheme 4.24

and it must be reduced earlier to the lactol. Use of fumarate nitroalkene was essential due to the instability of the stereoisomeric maleate, which resulted in wrong stereochemistry at C-6 in the final target molecule. The correct configuration at this stereocenter was obtained by a Mitsunobu reaction, while other simple functional group reactions allowed an overall formation of (-)-rosmarinecine.

Many interesting variants of this sequence were reported by Denmark et al. [53–55]. For instance, an intramolecular [3+2]/intermolecular [4+2]-cycloaddition cascade was utilized to synthesize a number of natural products, like (-)-rosmarinecine [52]. For this specific natural product, the Lewis acid-promoted reaction of nitroalkene with chiral enol ether afforded nitrosoacetal in excellent stereoselectivity (25:1, exo:endo) and 94% yield (Scheme 4.24). The lactol was obtained in 91% yield by the reduction of lactone functionality. The bicyclic lactam was obtained in 64% yield from lactol with Raney-Ni under hydrogen (160 psi). The chiral auxiliary was recovered in 98% yield. The protection of lactol followed by treatment with p-NO$_2$C$_6$H$_4$COOH, diethyl azodicarboxylate, and triphenylphosphine provided benzoate ester in 69% yield from bicyclic lactam. Finally, the deprotection of lactol in benzoate ester followed by exposure to Red-Al afforded rosmarinecine in 57% yield for the two-step method [10].

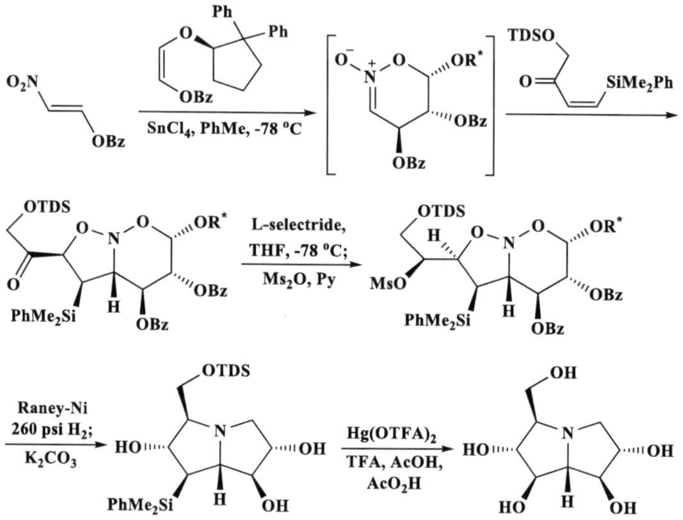

Scheme 4.25

This tandem intermolecular [4+2]/intermolecular [3+2]-cycloaddition approach was useful for the formation of (+)-casuarine. The nitroalkene was reacted with enol ether using stannous chloride at -78°C to provide an intermediate nitronate (Scheme 4.25) [56]. The oxazole-oxazine was obtained in 76% yield as a mixture dominating in the stereoisomer by a dipolar cycloaddition of nitronate to enone. The stereoselective reduction of ketone functionality in oxazole-oxazine followed by transformation to mesylate took place in 84% yield. The pyrrolizidine was obtained with 98% enantiomeric excess and in 64% yield by exposure of mesylate to Raney-Ni under high pressure. The (+)-casuarine was obtained in 84% yield by oxidative elimination of silyl group [10].

The tricyclic compounds were reduced to provide the beneficial polyfunctional starting compounds for the overall formation. The nitrogen-oxygen bonds proved much more rough than expected and comparatively severe conditions were mandatory before any reduction was reported to occur. The tricyclic compounds (nitroso acetals) were reduced to bicyclic lactams utilizing Ra-Ni/potassium fluoride as catalyst (Scheme 4.26) [17,18]. The reduction of tricyclic compound, obtained from *p*-nitrostyrene, was slightly unpredictable, and in some cases the secondary amine was obtained. The secondary amine was transformed to lactam with K_2CO_3 in MeOH [19,57].

Although the hydrogenation of nitroso acetals proceeded to expected product, this was not always the case, particularly when the nitroso acetal

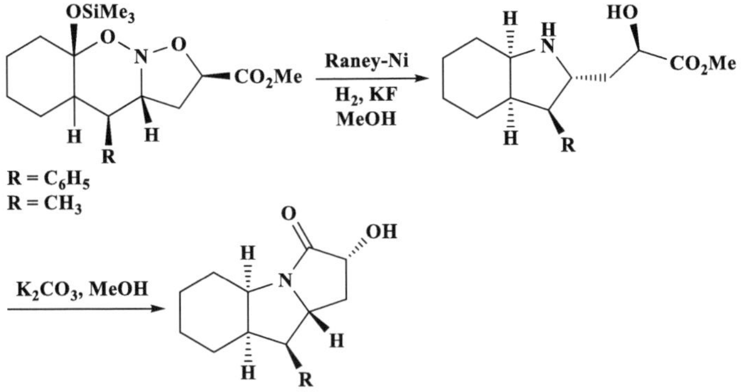

Scheme 4.26

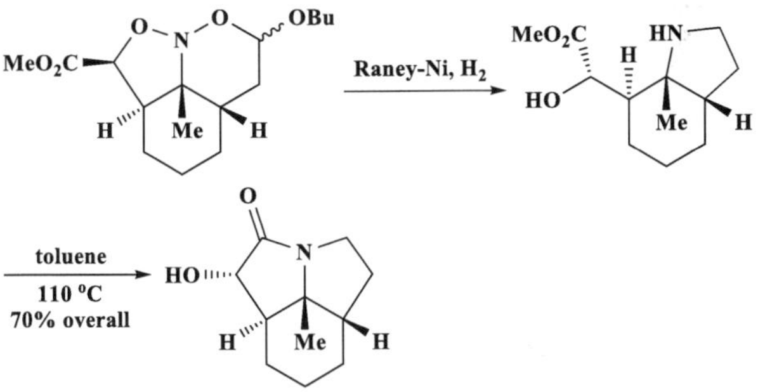

Scheme 4.27

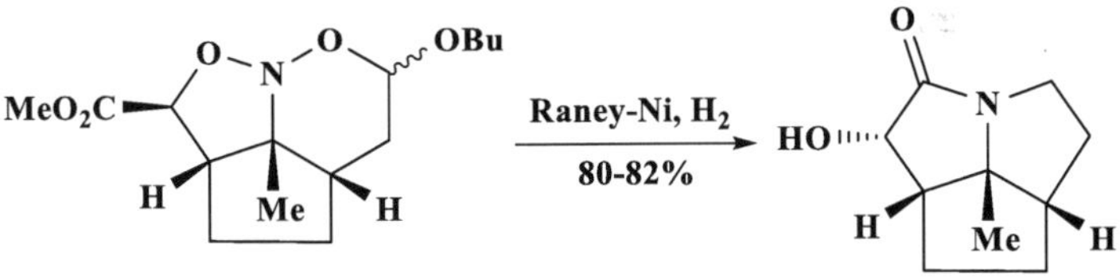

Scheme 4.28

was fused to another ring. This difficulty was first observed in early studies of the hydrogenation of oxazole-oxazine where the nitroso acetal was *trans*-fused to six-membered carbocycle (Scheme 4.27) [58]. The hydrogenation of nitroso acetal (with a *cis* fusion to a five-membered ring) proceeded uneventfully to tricyclic lactam at atmospheric pressure (Scheme 4.28). However, similar treatment of oxazole-oxazine left the oxazine ring intact. Higher H_2 pressures (160 psi) were essential to cleave both nitrogen-oxygen bonds to generate an intermediate where acylation not occurred. The transformation to lactam needed heating for 42 h in refluxing toluene. The difference in reactivity was credited to the construction of a *trans*-ring fusion in the case of three-atom chain [8].

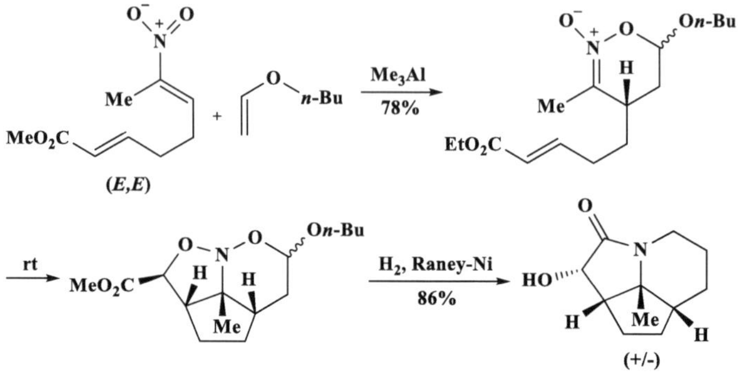

Scheme 4.29

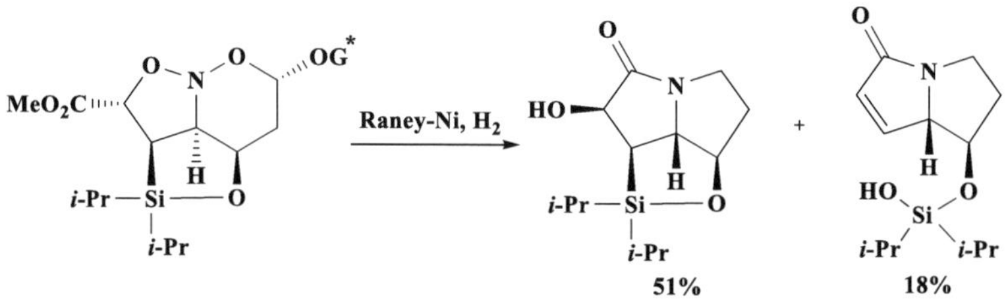

Scheme 4.30

The foundation of preparation of library was an employment of tandem cycloaddition of nitroalkenes to cast the polycyclic frameworks that will act as parallel formation frameworks. The known α-OH-lactam was obtained in racemic form by tandem [3+2]/[4+2]-cycloaddition of nitroalkene (E,E)-nitrone with n-butyl compound through the intermediacy of nitroso acetal and nitronate (Scheme 4.29). The reductive hydrogenolysis provided tricyclic lactam in excellent overall yield (86% over three steps) [59].

In the formation of detoxinine, the hydrogenation of an intermediate occurred to provide the lactam accompanied by unsaturated by-product (Scheme 4.30). The ratio of lactam to unsaturated by-product was independent of different reaction conditions. The side-product was supposed to be the outcome of a Peterson-type olefination [60]. The tricyclic lactam was resubjected to hydrogenation conditions and it was found that the olefination not took place after lactam synthesis. Thus, the elimination became a competing procedure during hydrogenation. Here again, the fused ring (this time a siloxane) hampered the acylation event and delivered an unwanted route of destruction [8].

An intramolecular alkylation was needed for the second ring-closure. Usually, alkylations (particularly with secondary centers) proceed slowly in

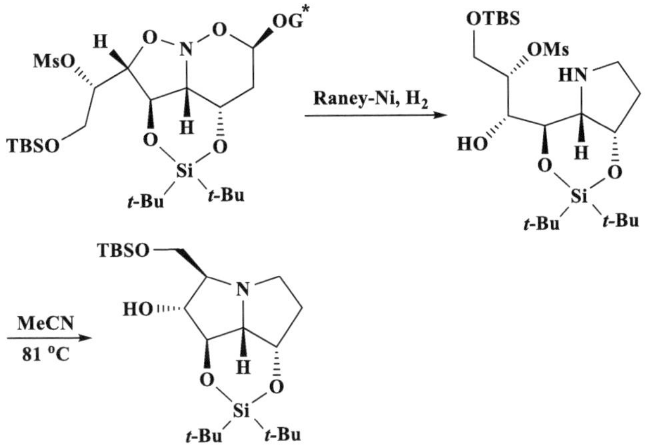

Scheme 4.31

comparison to acylations, as reported in the formation of australine (Scheme 4.31) [61]. The intermediate was obtained by hydrogenation of nitroso acetal, where a six-membered silyl acetal was embedded. This intermediate did not access a secondary reaction route, and pyrrolizidine ring-closure was achieved on heating in refluxing MeCN for 16 h [8].

The nitrocyclopentene derivatives are beneficial substrates in [4+2]-cycloadditions with enol ethers. Only one diastereomer of formed couple of bicyclic nitronates underwent a spontaneous [3+2] reaction to provide the nitroso acetal (Scheme 4.32) [62]. The 1-azafenestrane having all cis-configuration at the ring fusions was provided by reductive rearrangement and functional group transformations on this derivative.

First, the ring opened lactam was obtained in 59% yield by hydrogenation of lactam with Raney-Ni (Scheme 4.33). Similar to the hydrogenolysis of nitroso acetals, this reaction involved three distinct steps: reductive nitrogen–oxygen bond breakage, hemiacetal cleavage, and reduction of the aldehyde. Having obtained ring opened lactam, transformation of alcohol to a leaving group, followed by an intramolecular nucleophilic substitution was left to complete the synthesis. The tosylate was obtained in excellent yield by the reaction of alcohol with p-TsCl and Et$_3$N. The ring-closure occurred by displacement of p-toluenesulfonyl ester by the nucleophilic attack of lactam nitrogen. Whereas reaction with Et$_3$N in warm CHCl$_3$ not gave any reaction, reaction with t-BuOK provided lactam in 63% and 18% yield, respectively [63]. The reduction of β-lactam with a series of other reducing reagents was examined after the hydrogenolysis of lactam with Raney-Ni [43].

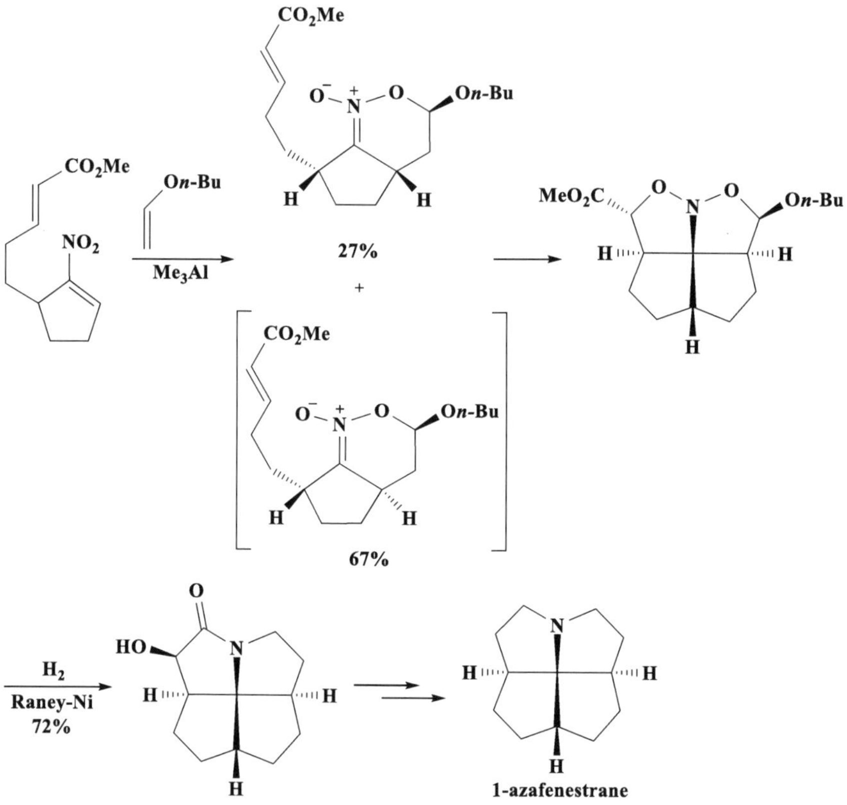

Scheme 4.32

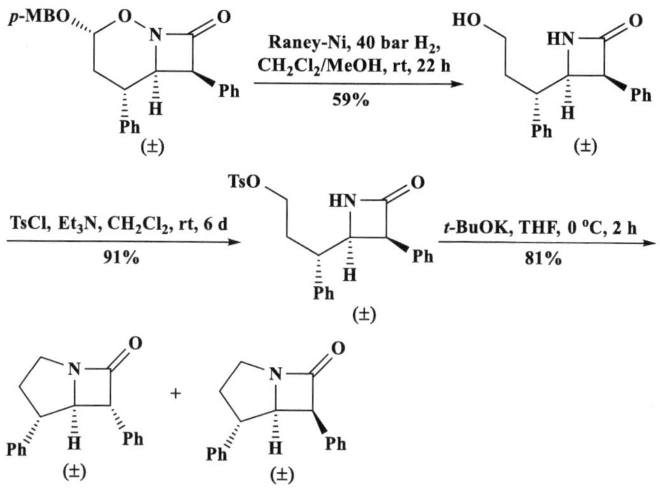

Scheme 4.33

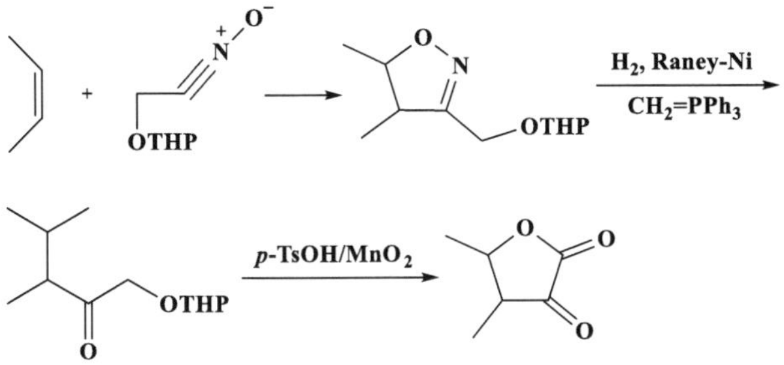

Scheme 4.34

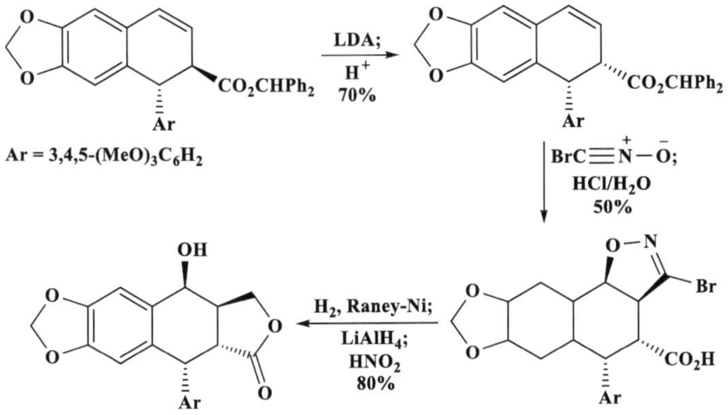

Scheme 4.35

4.5 Synthesis of five-membered *O*-heterocycles from oxazoles and oxazines

The isoxazoline product derived from the cycloaddition of nitrile oxide was cleaved by hydrogenolysis using Raney-Ni in a mixture of CH_3COOH, MeOH, and H_2O (1:8:2) [64]. The formed β-OH ketone acted as a substrate for the formation of 2-methylene lactone, which is an important subunit of various natural products (Scheme 4.34).

The 1,2-*trans*-stereochemistry was afforded when stereocontrolling elements were absent. The final transformation to podophyllotoxin still needed unfavorable lactone epimerization, whilst the subsequent steps of the synthesis of picropodophyllone increased. This challenging reaction was addressed by performing an isomerization at an earlier phase prior to an addition of C3-substituent or following hydrolysis to keto-acid. The former methodology enabled the stereocontrolled incorporation of C-3 hydroxymethyl group as realized in the work of Kaneko and Wong [65] and Vyas et al. [66] (Scheme 4.35).

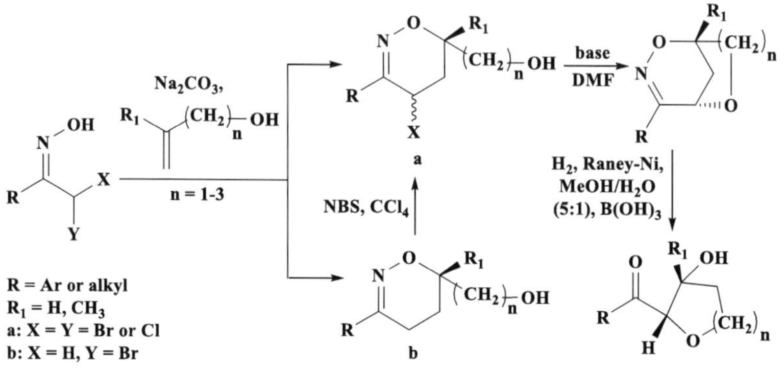

Scheme 4.36

The dihydro-4H-oxazinylmethanols were prepared by hetero-Diels-Alder reaction of nitrosoalkenes, produced in situ from halooximes, with allylic alcohols [67]. An incorporation of halogen atom at the 4-position of oxazine ring derived from halooximes afforded 2,6-dioxa-3-azabicycloalkenes through an intramolecular nucleophilic substitution of a halogen atom by OH group (Scheme 4.36). The reductive cleavage at the nitrogen-oxygen bond of oxazine ring afforded cyclic ether. The dihalooximes were selected as a starting compound to afford the halo-substituted oxazine derivatives. The dihaloketones were reacted with NH_2OH in methanol at rt for 2–4 days to provide the dihalooximes. The halovinylnitroso compounds, produced in situ by the treatment of dihalooximes with sodium carbonate or cesium carbonate, underwent a [4+2]-cycloaddition with allylic alcohols to provide the isomeric mixtures of 5,6-dihydro-4-halo-1,2-oxazines in 35–94% yield. A slight modification of reaction route was adopted. The bromo compounds were synthesized by bromination of compounds, which were prepared from monobromooximes, with N-bromosuccinimide. The 1,4-disubstituted 2,6-dioxa-3-azabicyclo[3.n+1.1]3-alkene was formed as single *trans*-isomer in 73–92% yield when these oxazines were reacted with a base like sodium hydride or potassium hydride. The reductive cleavage of nitrogen-oxygen bond of bicyclic oxazines with Raney-Ni (MeOH/water = 5:1) furnished acylated cyclic ethers stereoselectively in good yield [68,69].

4.6 Synthesis of six-membered heterocycles from oxazoles and oxazines

The protection of primary OH in starting compound as *t*-butyldimethylsilyl ether followed by reductive elimination of oxazolidinone gave primary

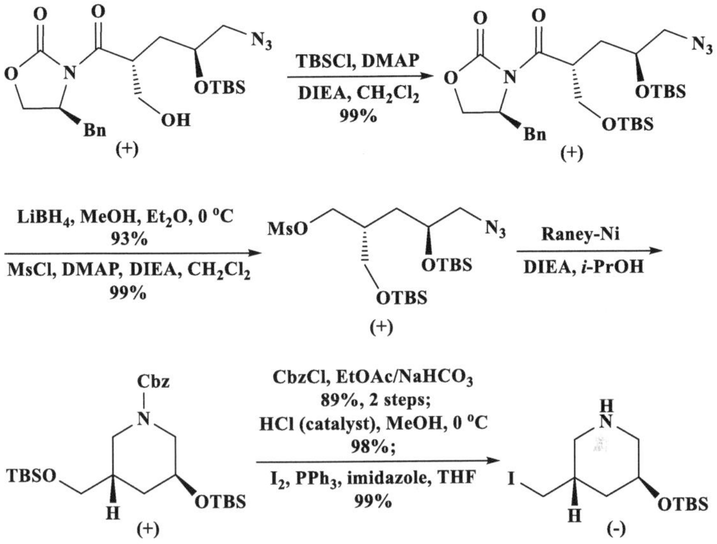

Scheme 4.37

alcohol, which was subsequently transformed to mesylate (Scheme 4.37). Hydrogenation of azide led to in situ cyclization of amine to provide the piperidine. The protection of amino group as Cbz carbamate, acidic hydrolysis of primary *t*-butyldimethylsilyl ether, and transformation of obtained alcohol to iodide completed the synthesis of desired product in 61% overall yield from known acid. This sequence proved to be both scalable and reliable [70–73].

Raney nickel- and palladium/carbon-catalyzed hydrogenation of different Baylis–Hillman adducts derived from activated alkenes and isoxazole-5-carboxaldehydes was reported by Saxena et al. [74]. They reported that the hydrogenation of Baylis–Hillman adducts with Raney nickel afforded *syn*-enaminones (through the opening of isoxazole ring) as a major product. The presence of H_3BO_3 in the reaction led to better *syn*-selectivity. Analogous hydrogenation of Baylis–Hillman adducts derived from isoxazole-4-carboxaldehyde derivatives utilizing palladium/carbon provided *anti*-isomer, without cleavage of the isoxazole ring, while hydrogenation with Raney nickel delivered decomposed products accompanied by pyridine derivatives as minor products (Scheme 4.38) [6].

The rearrangement of furfuryl amines has a wide range of applications in organic synthesis with its use being found in Ciufolini's [75,76] formation of desoxoprosopinine among various other instances from his group (Scheme 4.39).

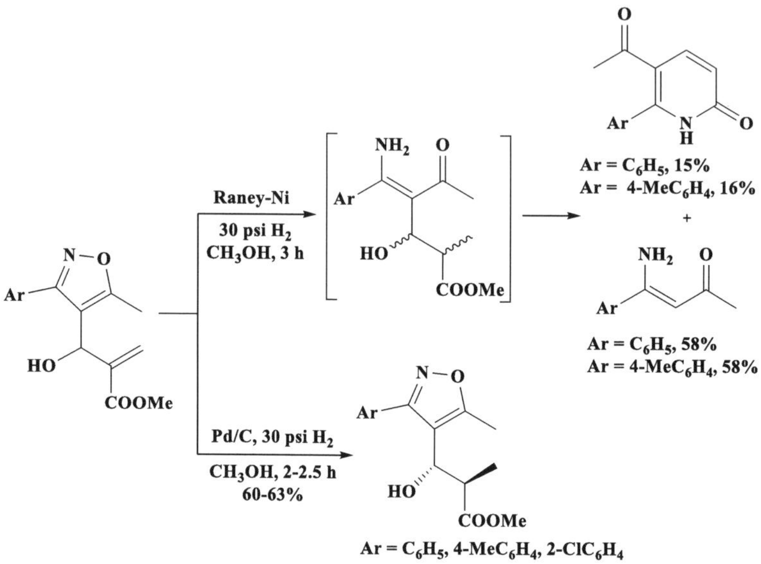

Scheme 4.38

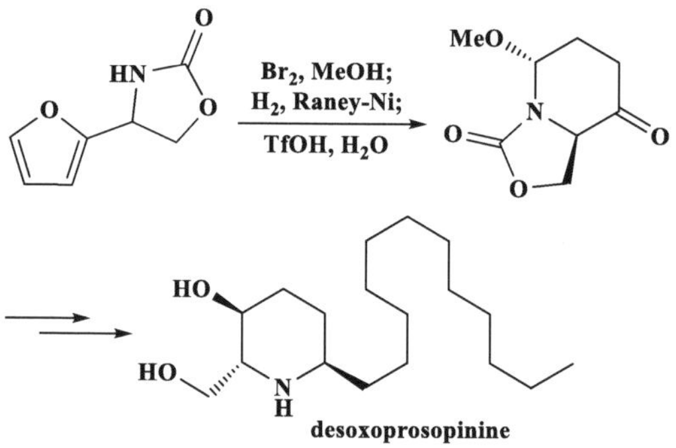

Scheme 4.39

Denmark and Gomez [77] described a tandem intramolecular [4+2]/intramolecular [3+2]-cycloaddition of nitroalkenes (Schemes 4.40,4.41). The nitronate was obtained when nitrone was reacted with stannous chloride. Heating the crude reaction mixture consisting of nitronate (n = 1) in toluene at 80°C for 90 minutes provided intermediate (n = 1) as a single diastereomer in 82% overall yield. The nitronate (n = 2) needed heating at 100°C in toluene for 3 days to provide the intermediate (n = 1) as a single diastereomer but in only 44% yield. The reduction of intermediate (n = 0, 1) with Raney-Ni under H$_2$ atmosphere (160 psi) delivered fused tricycles in 71% and 78% yield, respectively. The selectivity of this tandem

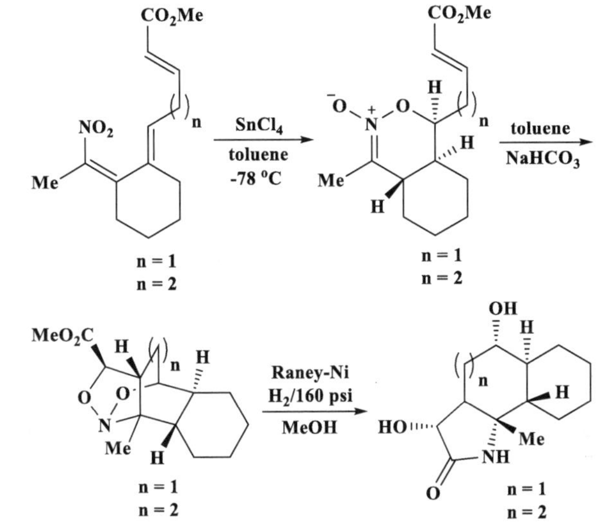

Scheme 4.40

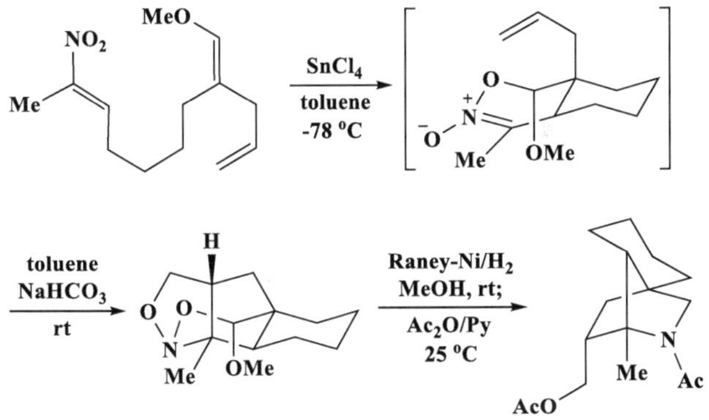

Scheme 4.41

sequence was remarkable in that fused tricycles each possess six contiguous stereogenic centers. Likewise, the nitroalkene was exposed to a Lewis acid to afford an intermediate in 87% yield. The reaction of intermediate with Raney-Ni and H_2 afforded bridged tricycle in 81% yield [10].

The synthesis of *Lycopodium* alkaloid cermizine D was published by Nishikawa et al. [78]; it has significant structural resemblances to the eastern portion of himeradine A (Scheme 4.42). The key step used citronellal derivative, and the aldehyde was exposed to organocatalyzed reductive Mannich cascade to synthesize the hydrazine oxazolidinone. The first ring of the quinolizidine portion of the natural product was constructed by

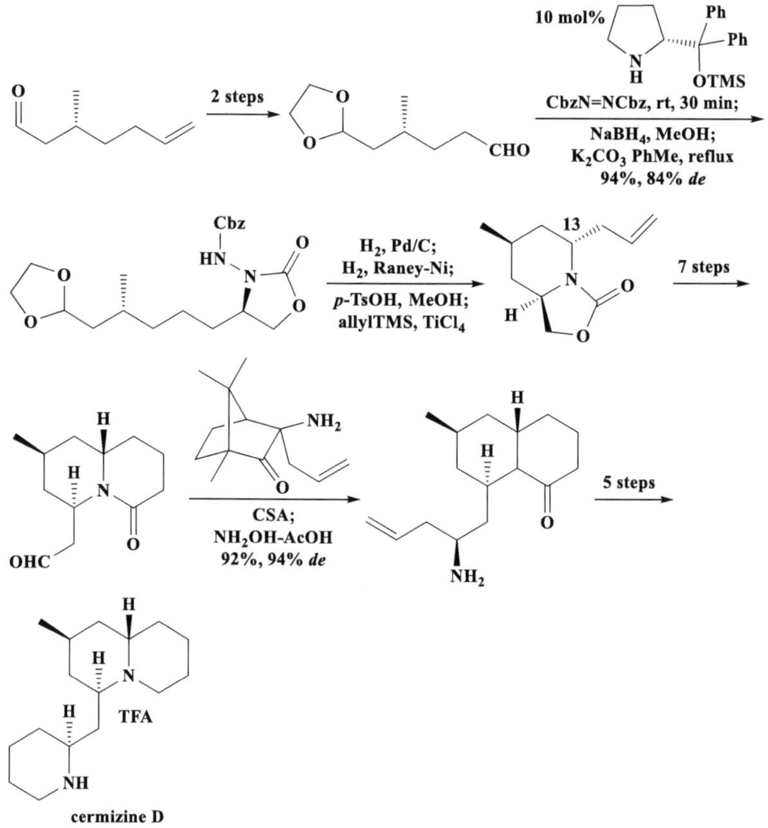

Scheme 4.42

a reaction sequence reduction/reduction/iminium formation/Sakurai to provide the bicyclic oxazolidinone. The second ring of quinolizidine was constructed by ring-closing metathesis as part of a seven-step sequence to afford the lactam aldehyde. The aldehyde was transformed to primary amine with amino allylated reagent. Another ring-closing metathesis reaction formed the remaining ring and a global reduction to synthesize the cermizine D.

This method needed 18 steps and started with carbonyl protection of (+)-citronellal, followed by a ruthenium-catalyzed oxidative cleavage of olefin bond (Scheme 4.43). The aldehyde was exposed to amination and subsequent reduction to provide the hydrazino alcohol. The oxazolidinone was formed after the reaction of hydrazino alcohol with potassium carbonate and reductive nitrogen–nitrogen bond cleavage of hydrazine group in

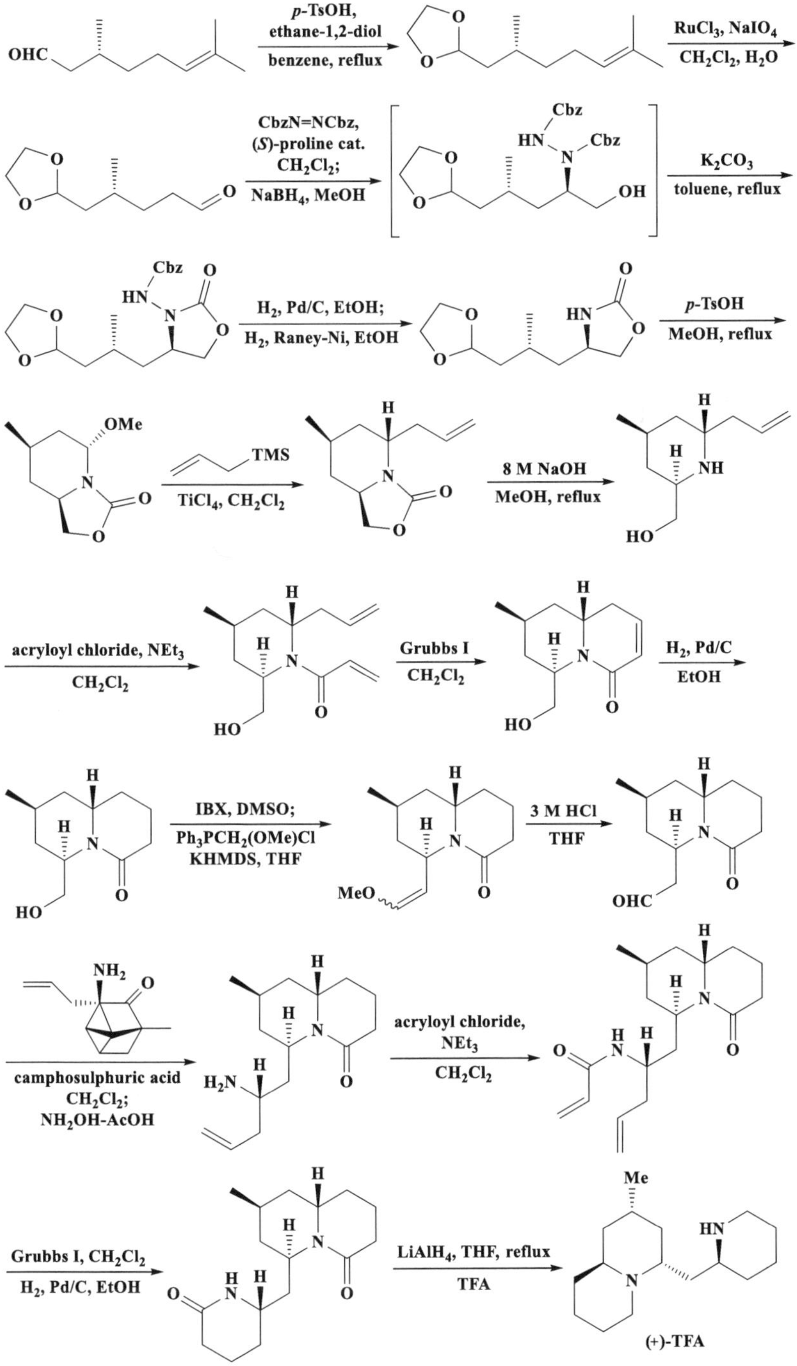

OHC
p-TsOH,
ethane-1,2-diol
benzene, reflux
O
O
RuCl3, NaIO4
CH2Cl2, H2O
CbzN=NCbz,
(S)-proline cat.
CH2Cl2;
NaBH4, MeOH
Cbz
HN
N
Cbz
OH
K2CO3
toluene, reflux
H2, Pd/C, EtOH;
H2, Raney-Ni, EtOH
HN
p-TsOH
MeOH, reflux
OMe
TMS
TiCl4, CH2Cl2
8 M NaOH
MeOH, reflux
NH
HO
acryloyl chloride, NEt3
CH2Cl2
Grubbs I
CH2Cl2
H2, Pd/C
EtOH
IBX, DMSO;
Ph3PCH2(OMe)Cl
KHMDS, THF
MeO
3 M HCl
THF
OHC
NH2
camphosulphuric acid
CH2Cl2;
NH2OH-AcOH
H2N
acryloyl chloride,
NEt3
CH2Cl2
Grubbs I, CH2Cl2
H2, Pd/C, EtOH
LiAlH4, THF, reflux
TFA
Me
HN
(+)-TFA

Scheme 4.43

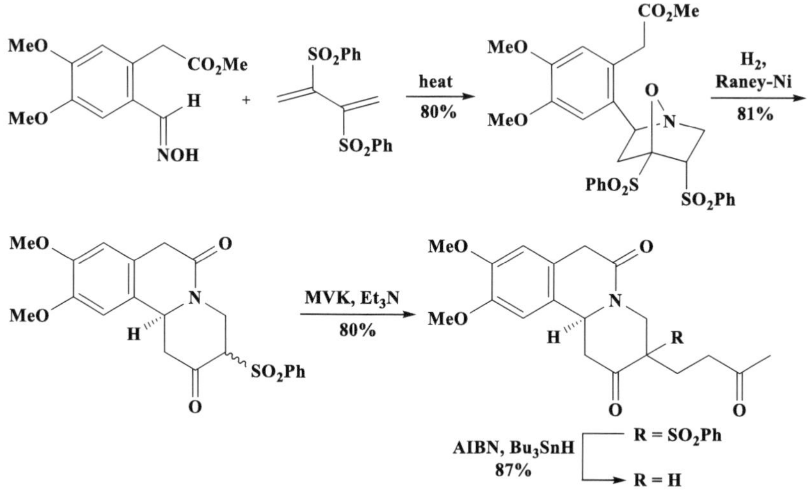

Scheme 4.44

oxazole. The acid-catalyzed cyclization led to aminoacetal, which afforded oxazole-oxazine exclusively under Hosomi-Sakurai allylation [79,80] conditions. After hydrolysis of the oxazolidinone, a three-step sequence including acroylation, ring-closing metathesis, and hydrogenation provided alcohol. This alcohol was oxidized with 2-iodoxybenzoic acid, and the aldehyde underwent a Wittig reaction [81,82], and acid hydrolysis. Kobayashi's aminoallylation [83] procedure was utilized to afford the homoallylamine. Finally, the piperidone was prepared utilizing same three-step sequence which was utilized to obtain the alcohol, starting with the acryloylation of quinolizine, followed by ring-closing metathesis and hydrogenation. The reduction of piperidin-2-ylmethyl-substituted quinolizine with lithium aluminum hydride gave target product.

The oxime was easily accessible from commercial 2-(3,4-dimethoxypheny) acetic acid and was reacted with phenylsulfonyldiene to synthesize the cycloadduct in 80% yield. The aryl group preferred to exist in an *endo*-orientation in the newly produced cycloadduct (Scheme 4.44). The cycloadduct was reduced with Raney-Ni under H_2 atmosphere to activate a cyclization cascade sequence whereby cleavage of nitrogen-oxygen bond was followed by spontaneous intramolecular acylation of nitrogen atom with proximal ester group to provide the quinolizine as a 1:1 mixture of diastereomers in excellent yield. The Robinson annulation reaction [84] depends on a specific order of operations to obtain

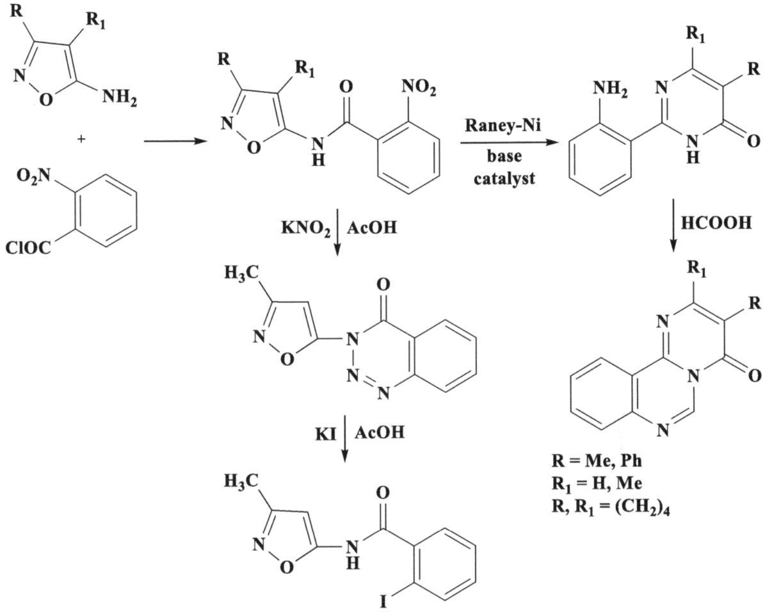

Scheme 4.45

R	R$_1$	yield (%)
C$_6$H$_5$	H	68
CH$_3$	H	65
C$_6$H$_5$	CH$_3$	70
(CH$_2$)$_4$	(CH$_2$)$_4$	69

the synthetically beneficial amounts of needed late-stage intermediates. The conjugate addition of MVK (methyl vinyl ketone) to quinolizine using catalytic amount of Et$_3$N gave a 1:1 mixture of diastereomeric ketones (R = SO$_2$Ph) in 80% yield [85]. The Sn-assisted phenylsulfonyl reduction (R = SO$_2$Ph) provided a compound (R = H) as a 1:1 mixture of diastereomeric amides [86].

The aminoisoxazole derivatives were reacted with acid chloride to provide the benzamides, which underwent a hydrogenolysis in the presence of a base to furnish the 2-(o-aminophenyl)-6-substituted pyrimidin-4(3H)-one. The pyrimido[1,2-c]quinazolin-4-ones were prepared by the reaction of 2-(o-aminophenyl)-6-substituted pyrimidin-4(3H)-one with HCOOH [87]. The benzamides were treated with KNO$_2$ in AcOH to give the benzotriazinone. The benzotriazinone was refluxed in AcOH with KI to provide the 2-iodo-N-isoxazolybenzamides (Scheme 4.45) [88,89].

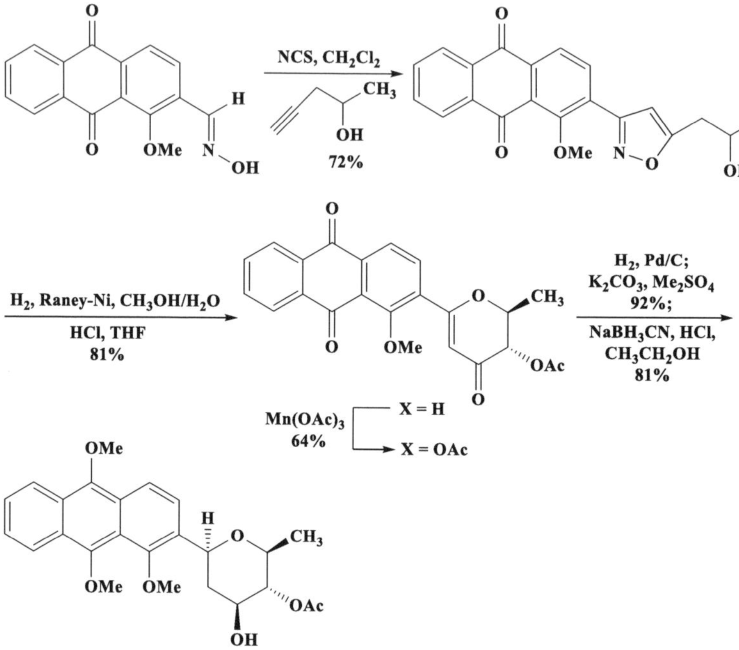

Scheme 4.46

The carbohydrate backbone was prepared by Hauser and Hu [90] from a cycloadduct, which was obtained by a dipolar cycloaddition. The nitrile oxide, which was synthesized by the treatment of anthraquinone oxime with *N*-chlorosuccinimide, underwent a cycloaddition with racemic alkyne to give the isoxazole. The pyranone was synthesized by hydrogenolysis of isoxazole followed by cyclization/dehydration of an intermediate diketone upon acid treatment. A mixture (3:1) of C2-epimers was obtained by α-acetoxylation of pyranone with manganic acetate. The 2-deoxy-*C*-anthracyl glycoside was formed in 81% yield as a single diastereomer when the major isomer was reduced under Tius's conditions [91] (Scheme 4.46).

The cycloaddition of nitrile oxide to olefin afforded isoxazoline as an inconsequential mixture of isomers when chloramine-T was added to a solution of starting compounds (Scheme 4.47) [92]. The cycloaddition reaction was incomplete, unless an excess of olefin was used, which led to an isolation of oximoyl chloride. This material was resubmitted to increase the yield by avoiding the incomplete transformation upon using an excess of volatile olefin.

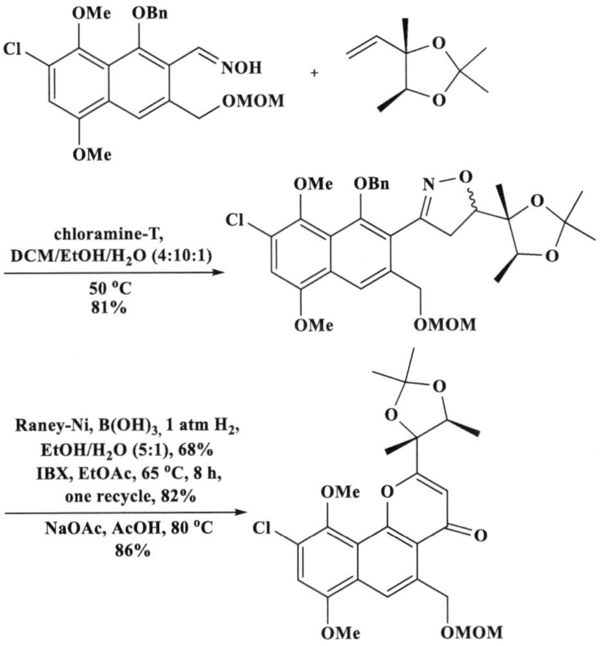

Scheme 4.47

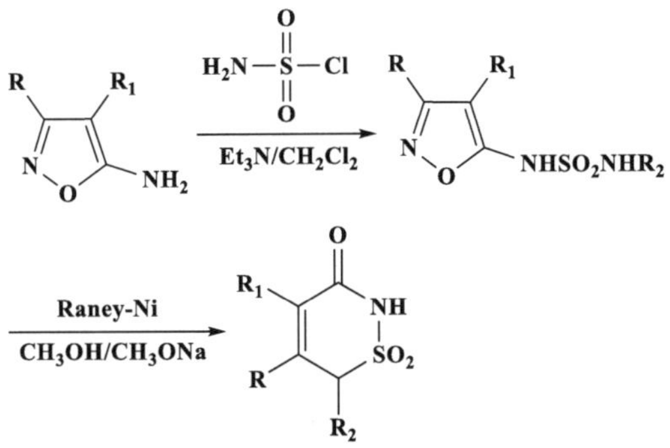

R	R_1	R_2	yield (%)
H	Me	H	47
Me	H	Et	20
Me	H	4-MeOC$_6$H$_5$	37

Scheme 4.48

The 4-methyl-5-isoxazolamine was reacted with sulfamoyl chloride in CH_2Cl_2/Et_3N to provide the sulfamoyl isoxazolamine [93]. Hydrogenation of sulfamoyl isoxazolamine in CH_3OH over Raney-Ni catalyst in CH_3ONa provided final product (Scheme 4.48) [89].

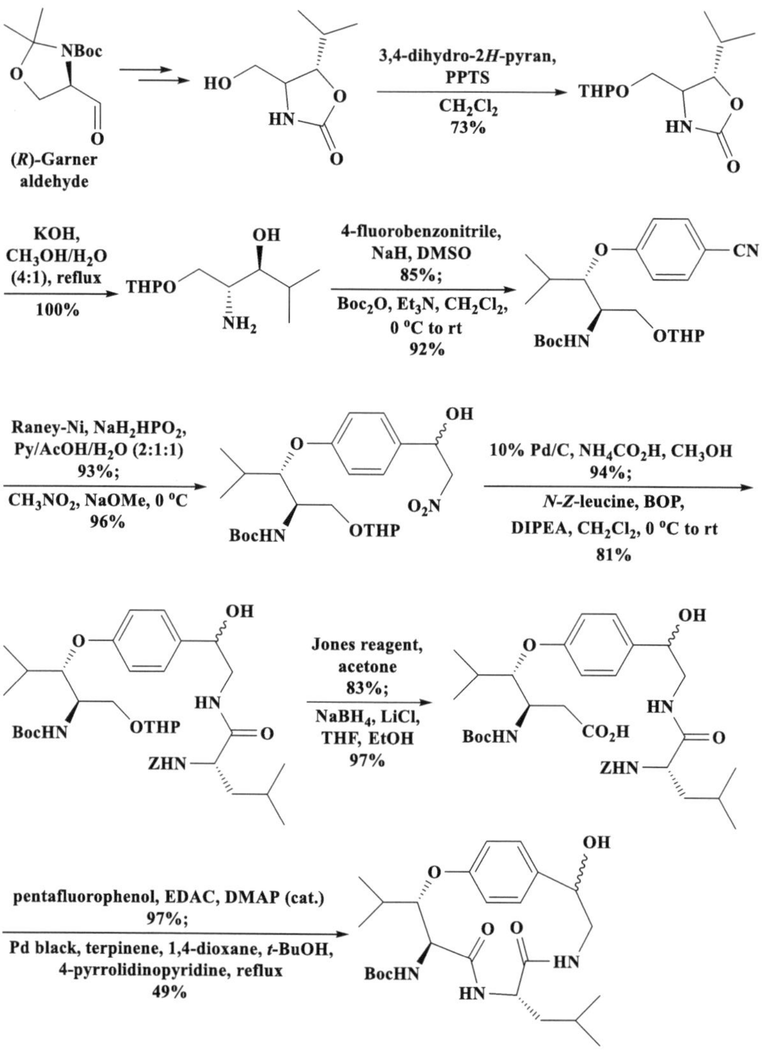

Scheme 4.49

4.7 Synthesis of higher-membered heterocycles from oxazoles and oxazines

The 14-membered cyclopeptides are the most synthetically challenging goals due to the strain associated with the ring size. An alkyl-aryl ether linkage was present in their structures. East and Joullie [94] reported a direct synthesis of cyclopeptide starting from D-Garner aldehyde. Garner aldehyde was transformed to oxazolidine following a literature process. Further, this oxazolidine was transformed to 3-hydroxyleucine derivative, which was reacted with 4-fluorobenzonitrile followed by Boc-protection to synthesize the cyano derivative. Raney nickel-assisted reduction of cyanide to aldehyde followed by treatment with CH_3NO_2 under Henry conditions delivered NO_2 derivative, which was consequently reduced and reacted with leucine derivative to produce an intermediate. The obtained intermediate was then converted into acid under Jones oxidation conditions and lithium borohydride-assisted selective reduction of the benzylic ketone and finally cyclization followed by deprotection allowed the completion of the formation of cyclopeptide (Scheme 4.49).

References

[1] (a) Kaur N, Grewal P, Bhardwaj P, Devi M, Ahlawat N, Verma Y. Synthesis of five-membered N-heterocycles using silver metal. Synth Commun 2020;49:3058–100; (b) Kaur N, Devi M, Verma Y, Grewal P, Jangid NK, Dwivedi J. Seven and higher-membered oxygen heterocycles: metal and non-metal. Synth Commun 2019;49:1508–42; (c) Kaur N. Ionic liquid promoted eco-friendly and efficient synthesis of six-membered N-polyheterocycles. Curr Org Synth 2018;15:1124–46; (d) Kaur N, Ahlawat N, Bhardwaj P, Verma Y, Grewal P, Jangid NK. Ag-mediated synthesis of six-membered N-heterocycles. Synth Commun 2020;50:753–95; (e) Kaur N. Ultrasound assisted synthesis of six-membered N-heterocycles. Mini Rev Org Chem 2018;15:520–36; (f) Kaur N, Verma Y, Ahlawat N, Grewal P, Bhardwaj P, Jangid NK. Copper-assisted synthesis of five-membered O-heterocycles. Inorg Nano Met Chem 2020;50:705–40.

[2] (a) Kaur N, Ahlawat N, Verma Y, Grewal P, Bhardwaj P, Jangid NK. Cu-Assisted C-N bond formations in six-membered N-heterocycle synthesis. Synth Commun 2020;50:1075–132; (b) Kaur N, Verma Y, Grewal P, Ahlawat N, Bhardwaj P, Jangid NK. Palladium acetate assisted synthesis of five-membered N-polyheterocycles. Synth Commun 2020;50:1567–621; (c) Kaur N, Verma Y, Grewal P, Ahlawat N, Bhardwaj P, Jangid NK. Photochemical C-N bond forming reactions for the synthesis of five-membered fused N-heterocycles. Synth Commun 2020;50:1286–334; (d) Kaur N, Ahlawat N, Verma Y, Grewal P, Bhardwaj P, Jangid NK. Metal and organo-complex promoted synthesis of fused five-membered O-heterocycles. Synth Commun 2020;50:457–505; (e) Kaur N, Ahlawat N, Bhardwaj P, Verma Y, Grewal P, Jangid NK. Synthesis of five-membered N-heterocycles using Rh based metal catalysts. Synth Commun 2020;50:137–60; (f) Kaur N, Kishore D. Synthesis of 2-(oxadiazolo, pyrimido, imidazolo, and benzimidazolo) substituted analogues of 1,4-benzodiazepin-5-carboxamides linked through a phenoxyl bridge. J Chem Sci 2014;126:1861–7; (g) Sharma P, Kaur N, Sirohi R, Kishore D. Microwave assisted facile one pot synthesis of novel 5-carboxamido substituted analogues of 1,4-benzodiazepin-2-one of medicinal interest. Bull Chem Soc Ethiop 2013;27:301–7.

[3] (a) Kaur N, Devi M, Verma Y, Grewal P, Bhardwaj P, Ahlawat N, Jangid NK. Photochemical synthesis of fused five-membered O-heterocycles. Curr Green Chem 2019;6:155–83; (b) Kaur N, Devi M, Verma Y, Grewal P, Bhardwaj P, Ahlawat N, Jangid NK. Applications of metal and non-metal catalysts for the synthesis of oxygen containing five-membered polyheterocycles: a mini review. SN Appl Sci 2019;1:1–32; (c) Kaur N, Ahlawat N, Verma Y, Bhardwaj P, Grewal P, Jangid NK. Rhodium

catalysis in the synthesis of fused five-membered N-heterocycles. Inorg Nano Met Chem 2020;50:1260–89; (d) Kaur N. Synthesis of three-membered and four-membered heterocycles with the assistance of photochemical reactions. J Heterocycl Chem 2019;56:1141–67; (e) Kaur N, Ahlawat N, Grewal P, Bhardwaj P, Verma Y. Organo or metal complex catalyzed synthesis of five-membered oxygen heterocycles. Curr Org Chem 2019;23:2822–47; (f) Kaur N, Tyagi R, Kishore D. Expedient protocols for the installation of 1,5-benzoazepino-based privileged templates on the 2-position of 1,4-benzodiazepine through a phenoxyl spacer. J Heterocycl Chem 2014;51:E340–3; (g) Kaur N. Recent trends in the chemistry of privileged scaffold: 1,4-benzodiazepine. Int J Pharm Biol Sci 2013;4:485–513.
[4] (a) Kaur N, Ahlawat N, Verma Y, Grewal P, Bhardwaj P. A review of ruthenium catalyzed C-N bond formation reactions for the synthesis of five-membered N-heterocycles. Curr Org Chem 2019;23:1901–44; (b) Kaur N, Bhardwaj P, Devi M, Verma Y, Grewal P. Gold-catalyzed C-O bond forming reactions for the synthesis of six-membered O-heterocycles. SN Appl Sci 2019;1:1–37; (c) Kaur N. Ionic liquid assisted synthesis of six-membered oxygen heterocycles. SN Appl Sci 2019;1:1–20; (d) Kaur N, Jangid NK, Sharma V. Metal- and nonmetal-catalyzed synthesis of five-membered S,N-heterocycles. J Sulfur Chem 2018;39:193–236; (e) Kaur N, Jangid NK, Rawat V. Synthesis of heterocycles through platinum-catalyzed reactions. Curr Catal 2018;7:3–25; (f) Kaur N, Kishore D. Montmorillonite: an efficient, heterogeneous, and green catalyst for organic synthesis. J Chem Pharm Res 2012;4:991–1015; (g) Chauhan R, Kaur N, Rajendra SS, Dwivedi J. Application of chalcone in synthesis of 1-(1,5-benzodiazepino) substituted analogues of indole. Rasayan J Chem 2015;8:115–22.
[5] Singh P, Samanta K, Das SK, Panda G. Amino acid chirons: a tool for asymmetric synthesis of heterocycles. Org Biomol Chem 2014;12:6297–339.
[6] Basavaiah D, Reddy BS, Badsara SS. Recent contributions from the Baylis-Hillman reaction to organic chemistry. Chem Rev 2010;110:5447–674.
[7] Pulz R, Watanabe T, Schade W, Reissig H-U. A stereodivergent synthesis of enantiopure 3-methoxypyrrolidines and 3-methoxy-2,5-dihydropyrroles from 3,6-dihydro-2H-1,2-oxazines. Synlett 2000;7:983–6.
[8] Denmark SE, Cottell JJ. Synthesis of (+)-1-epiaustraline. J Org Chem 2001;66:4276–84.
[9] Mitchinson A, Nadin A. Saturated nitrogen heterocycles. J Chem Soc, Perkin Trans 1 1999;18:2553–81.
[10] Padwa A, Bur SK. The domino way to heterocycles. Tetrahedron 2007;63:5341–78.
[11] Domling A. Recent developments in isocyanide based multicomponent reactions in applied chemistry. Chem Rev 2006;106:17–89.
[12] Domling A, Ugi I. Multicomponent reactions with isocyanides. Angew Chem, Int Ed 2000;39:3168–210.
[13] Tietze LF, Beifuss U. Sequential transformations in organic chemistry: a synthetic strategy with a future. Angew Chem, Int Ed 1993;32:131–63.
[14] Tietze LF. Domino reactions in organic synthesis. Chem Rev 1996;96:115–36.
[15] Abbadi A and van Bekkum H. 2001. Fine chemicals through heterogeneous catalysis. R.A. Sheldon and H. van Bekkum (Eds.). Wiley-VHC, 380.
[16] Denmark SE, Thorarensen A. The tandem cycloaddition chemistry of nitroalkenes. A novel synthesis of (-)-hastanecine. J Org Chem 1994;59:5672–80.
[17] Das NB, Torsell KBG. Silyl nitronates in organic synthesis. Tetrahedron 1983;39:2227–30.
[18] Mukerji SK, Torsell KBG. Silyl nitronates in organic synthesis. Silylation of secondary nitro compounds. Preparation of nitroso compounds and alpha,beta-unsaturated aldehydes. Acta Chem Scand Ser B 1981;35:643–8.

[19] Brenner M, Seebach D. Enantioselective preparation of γ-amino acids and γ-lactams from nitro olefins and carboxylic acids, with the valine-derived 4-isopropyl-5,5-diphenyl-1,3-oxazolidin-2-one as an auxiliary. Helv Chim Acta 1999;82:2365–79.
[20] Mulzer J, Zuhse R, Schmiechen R. Enantioselective synthesis of the antidepressant rolipram by Michael addition to a nitroolefin. Angew Chem, Int Ed 1992;31:870–2.
[21] Palomo C, Aizpurua JM, Oiarbide M, García JM, González A, Odriozola I, Linden A. Diastereoselective Michael reactions of (1R)-(+)-camphor methyl ketone enolates with nitro olefins. Tetrahedron Lett 2001;42:4829–31.
[22] Hoshiko T, Ishihara H, Shino M, Mori N. The synthesis of *trans*-3,4-bis-(3,4-dihydroxyphenyl)pyrrolidine, a novel DA1 agonist. Chem Pharm Bull 1993;41:633–5.
[23] Aslanian R, Lee G, Iyer RV, Shih N-Y, Piwinski JJ, Draper RW, McPhail AT. An asymmetric synthesis of the novel H3 agonist (+)-(3R,4R)-3-(4-imidazolyl)-4-methylpyrrolidine dihydrochloride (Sch 50971). Tetrahedron: Asymmetry 2000;11:3867–71.
[24] Nevalainen M, Kauppinen PM, Koskinen AM. Synthesis of Fmoc-protected *trans*-4-methylproline. J Org Chem 2001;66:2061–6.
[25] Qiu X, Qing F. Synthesis of Boc-protected *cis*- and *trans*-4-trifluoromethyl-D-prolines. J Chem Soc, Perkin Trans 1 2002;18:2052–7.
[26] Roy AK, Pathak R, Yadav GP, Maulik PR, Batra S. Neighboring-group effect: DBU-promoted ring transformation of substituted isoxazoles to substituted pyrroles. Synthesis 2006;6:1021–7.
[27] Singh V, Saxena R, Batra S. Simple and efficient synthesis of substituted 2-pyrrolidinones, 2-pyrrolones, and pyrrolidines from enaminones of Baylis-Hillman derivatives of 3-isoxazolecarbaldehydes. J Org Chem 2005;70:353–6.
[28] Cohen N, Banner BL, Blount JF, Weber G, Tsai M, Saucy G. Synthesis of novel spiro heterocycles. 2-Amino-7-oxa-3-thia-1-azaspiro[5.5]undec-1-enes. J Org Chem 1974;39:1824–33.
[29] Ghabrial SS, Thomsen I, Torssell KBG. Synthesis of biheteroaromatic compounds via the isoxazoline route. Acta Chem Scand B 1987;41:426–34.
[30] Cai X-C, Wu X, Snider BB. Synthesis of the spiroiminal moiety of marineosins A and B. Org Lett 2010;12:1600–3.
[31] Winchester B, Fleet GWJ. Amino-sugar glycosidase inhibitors: versatile tools for glycobiologists. Glycobiology 1992;2:199–210.
[32] Look GC, Fotsch CH, Wong C-H. Enzyme-catalyzed organic synthesis: practical routes to aza sugars and their analogs for use as glyco processing inhibitors. Acc Chem Res 1993;26:182–90.
[33] Ganem B. Inhibitors of carbohydrate-processing enzymes: design and synthesis of sugar-shaped heterocycles. Acc Chem Res 1996;29:340–7.
[34] Hümmer W, Dubois E, Gracza T, Jäger V. Halocyclization and palladium(II)-catalyzed amidocarbonylation of unsaturated aminopolyols. Synthesis of 1,4-iminoglycitols as potential glycosidase inhibitors. Synthesis 1997;6:634–42.
[35] Bols M. 1-Aza sugars, apparent transition state analogues of equatorial glycoside formation/cleavage. Acc Chem Res 1998;31:1–8.
[36] Hornuth S, Reissig H-U. Stereoselective synthesis of 3(2H)-dihydrofuranones by addition of lithiated methoxyallene to chiral aldehydes. J Org Chem 1994;59:67–73.
[37] Amombo MO, Hausherr A, Reissig H-U. An expedient synthesis of pyrrole derivatives by reaction of lithiated methoxyallenes with imines. Synlett 1999;12:1871–4.
[38] Beemelmanns C, Reissig HU. A short formal total synthesis of strychnine with a samarium diiodide induced cascade reaction as the key step. Angew Chem, Int Ed 2010;49:8021–5.

[39] Reissig H-U, Hormuth S, Schade W, Amombo MO, Watanabe T, Pulz R, Hausherr A, Zimmer R. Stereoselective synthesis of heterocycles with lithiated methoxyallene. J Heterocycl Chem 2000;31:597–606.
[40] Jäger V, Schröter P. Synthesis of amino sugars via isoxazolines: D-allosamine. Synthesis 1990;7:556–60.
[41] Müller R, Leibold T, Pätzel M, Jäger V. A new synthesis of 1,3,4-trideoxy-1,4-iminoglycitols of varying chain length by (C3+CN)-coupling of allyl halides with glycononitrile oxides. Angew Chem, Int Ed 1994;33:1295–8.
[42] Angermann J, Homann K, Reissig H-U, Zimmer R. Synthesis and *cis*-dihydroxylation of 6*H*-1,2-oxazines: synthesis of dihydroxyprolinols. Synlett 1995;10:1014–16.
[43] Denmark SE, Thorarensen A. Tandem [4+2]/[3+2] cycloadditions of nitroalkenes. Chem Rev 1996;96:137–66.
[44] Kuster GJT. Solution- and solid-phase synthesis of novel heterocycles via high-pressure promoted cycloadditions of nitroalkenes, Nijmegen: Radboud University; 2001. PhD Thesis.
[45] Gallos JK, Sarli VC, Varvogli AC, Papadoyanni CZ, Papaspyrou SD, Argyropoulos NG. The hetero-Diels-Alder addition of ethyl 2-nitrosoacrylate to electron-rich alkenes as a route to unnatural α-amino acids. Tetrahedron Lett 2003;44:3905–9.
[46] Ager DJ. The Peterson olefination reaction. Org React 1990;38:1–223.
[47] Xie M. Synthesis of nitrogen-containing heterocycles using nitro compounds as building blocks: Part I: Synthesis of 3-substituted azepanes. Part II: Synthesis of 1-azoniapropellanes as phase transfer catalysts. University of Illinois at Urbana-Champaign; 2010. PhD Thesis.
[48] Denmark SE, Herbert B. Synthesis of (1*R*,2*R*,3*R*,7*R*,7a*R*)-hexahydro-3-(hydroxymethyl)-1*H*-pyrrolizine-1,2,7-triol: 7-epiaustraline. J Am Chem Soc 1998;120:7357–8.
[49] Liddel JR. Pyrrolizidine alkaloids. Nat Prod Rep 2002;19:773–81.
[50] Denmark SE, Hurd AR. Tandem [4+2]/[3+2] cycloadditions with nitroethylene. J Org Chem 1998;63:3045–50.
[51] Denmark SE, Thorarensen A, Middleton DS. A general strategy for the synthesis of *cis*-substituted pyrrolizidine bases. The synthesis of (-)-rosmarinecine. J Org Chem 1995;60:3574–5.
[52] Denmark SE, Thorarensen A, Middleton DS. Tandem [4+2]/[3+2] cycloadditions of nitroalkenes. 9. Synthesis of (-)-rosmarinecine. J Am Chem Soc 1996;118:8266–77.
[53] Denmark SE, Middleton DS. Tandem inter [4+2]/intra [3+2] cycloadditions. 17. The spiro mode. Efficient and highly selective synthesis of azapropellanes. J Org Chem 1998;63:1604–18.
[54] Denmark SE, Throarensen A. Tandem [4+2]/[3+2] cycloadditions of nitroalkenes. 11. The synthesis of (+)-crotanecine. J Am Chem Soc 1997;119:125–37.
[55] Denmark SE, Dixon JA. Tandem inter [4+2]/intra [3+2] cycloadditions of nitroalkenes. A versatile asymmetric synthesis of highly functionalized aminocyclopentanes. J Org Chem 1997;62:7086–7.
[56] Denmark SE, Hurd AR. Synthesis of (+)-casuarine. J Org Chem 2000;65:2875–86.
[57] Brook MA, Eebach D. Cyclic nitronates from the diastereoselective addition of 1-trimethylsilyloxycyclohexene to nitroolefins. Starting materials for stereoselective Henry reactions and 1,3-dipolar cycloadditions. Can J Chem 1987;65:836–50.
[58] Denmark SE, Moon Y-C, Senanayake CBW. Tandem [4+2]/[3+2] cycloadditions: facile and stereoselective construction of polycyclic frameworks. J Am Chem Soc 1990;112:311–15.
[59] Wolf LM. Catalyst development in asymmetric phase transfer catalysis employing QSAR methods & computational investigations on the stereochemical course of the

addition of allylsilanes to aldehydes. University of Illinois at Urbana-Champaign; 2012. PhD Thesis.
[60] Righi P, Marotta E, Rosini G. Linear aminopolyhydroxylated structures through rapid domino assembly of a highly functionalized heterotricyclic system and its selective cleavage. Chem Eur J 1998;4:2501–12.
[61] Denmark SE, Martinborough EA. Enantioselective total synthesis of (+)-castanospermine, (+)-6-epicastanospermine, (+)-australine, and (+)-3-epiaustraline. J Am Chem Soc 1999;121:3046–56.
[62] Denmark SE, Montgomery JI, Kramps LA. Synthesis, X-ray crystallography, and computational analysis of 1-azafenestranes. J Am Chem Soc 2006;128:11620–30.
[63] Barrow KD, Spotswood TM. Stereochemistry and P.M.R. spectra of β-lactams. Tetrahedron Lett 1965;6:3325–35.
[64] Kozikowski AP, Ghose AK. The isoxazoline route to α-methylen lactones. Tetrahedron Lett 1993;24:2623–6.
[65] Kaneko T, Wong H. Total synthesis of (±) podophyllotoxin. Tetrahedron Lett 1987;28:517–20.
[66] Vyas DM, Skonezny PM, Jenkins TA, Doyle TW. Total synthesis of (±) epipodophyllotoxin via a (3+2)-cycloaddition strategy. Tetrahedron Lett 1986;27:3099–102.
[67] Stevensen TM, Patel KM, Crouse BA, Folgar MP, Hutchison CD, Pine KK. Oxazine ether herbicides. Synthesis and chemistry of agrochemicals. ACS Symp Ser 1995;584:197–205.
[68] Fisera L, Goljer I, Jaroskora L. Reaction of tetrahydrofuroisoxazoles with molybdenum hexacarbonyl. A new route to preparation of 3-substituted tetrahydro- and dihydrofuran derivatives. Collect Czech Chem Commun 1988;53:1753–60.
[69] Lee I-YC, Lee JH, Lee HW. Synthetic utilization of 2,6-dioxa-3-azabicycloalkenes toward cyclic ethers. Bull Korean Chem Soc 2002;23:537–8.
[70] Stork G, Benaim J. Monoalkylation of alpha,beta-unsaturated ketones via metalloenamines. J Am Chem Soc 1971;93:5938–9.
[71] Petersen JS, Tötenberg-Kaulen S, Rapoport H. Synthesis of (+-)-omega-aza[x.y.1]bicycloalkanes by an intramolecular Mannich reaction. J Org Chem 1984;49:2948–53.
[72] Panek JS, Beresis RT, Celatka CA. Studies directed toward the synthesis of ulapualide A. Asymmetric synthesis of the C26-C42 fragment. J Org Chem 1996;61:6494–5.
[73] Beshore DC, Smith AB. The lyconadins: enantioselective total synthesis of (+)-lyconadin A and (-)-lyconadin B. J Am Chem Soc 2008;130:13778–89.
[74] Saxena R, Singh V, Batra S. Studies on the catalytic hydrogenation of Baylis-Hillman derivatives of substituted isoxazolecarbaldehydes. Unusual retention of isoxazole ring during Pd-C-promoted hydrogenation of Baylis-Hillman adducts. Tetrahedron 2004;60:10311–20.
[75] Ciufolini MA, Wood CY. The aza-Achmatowicz rearrangement: a route to useful building blocks for N-containing structures. Tetrahedron Lett 1986;27:5085–8.
[76] Ciufolini MA, Hermann CW, Whitmire KH, Byrne NE. Chemoenzymatic preparation of *trans*-2,6-dialkylpiperidines and of other azacycle building blocks. Total synthesis of (+)-desoxoprosopinine. J Am Chem Soc 1989;111:3473–5.
[77] Denmark SE, Gomez L. Tandem double intramolecular [4+2]/[3+2] cycloadditions of nitroalkenes. Org Lett 2001;3:2907–10.
[78] Nishikawa Y, Kitajima M, Takayama H. First asymmetric total synthesis of cernuane-type *Lycopodium* alkaloids, cernuine, and cermizine D. Org Lett 2008;10:1987–90.
[79] Hosomi A, Sakurai H. Synthesis of γ,δ-unsaturated alcohols from allylsilanes and carbonyl compounds in the presence of titanium tetrachloride. Tetrahedron Lett 1976;17:1295–8.

[80] Hosomi A. Characteristics in the reactions of allylsilanes and their applications to versatile synthetic equivalents. Acc Chem Res 1988;21:200–6.
[81] Wittig G, Geissler G. Zur reaktionsweise des pentaphenyl-phosphors und einiger derivate. Liebigs Ann Chem 1953;580:44–57.
[82] Maryanoff BE, Reitz AB. The Wittig olefination reaction and modifications involving phosphoryl-stabilized carbanions. Stereochemistry, mechanism, and selected synthetic aspects. Chem Rev 1989;89:863–927.
[83] Sugiura M, Mori C, Kobayashi S. Enantioselective transfer aminoallylation: synthesis of optically active homoallylic primary amines. J Am Chem Soc 2006;128:11038–9.
[84] Gawley RH. The Robinson annelation and related reactions. Synthesis 2000;1976:777–94.
[85] Rosenmund P, Brandt B, Flecker P, Eberhard H. Stereoselektive total synthese vonrac-yohimb-15-enon. Liebigs Ann Chem 1990;9:857–62.
[86] Flick AC, Padwa A. A conjugate addition-dipolar cycloaddition approach towards the synthesis of various alkaloids. ARKIVOC 2011(vi):137–61.
[87] Plescia S, Ajello E, Sprio V, Marino ML. Synthesis of some pyrimido[1,2-c]quinazolin-4-one derivatives. J Heterocycl Chem 1974;11:603–6.
[88] Raffa D, Daidone G, Maggio B, Schillaci D, Plescia F, Torta L. Synthesis and antifungal activity of new N-isoxazolyl-2-iodobenzamides. Farmaco 1999;54:90–4.
[89] Hamama WS, Ibrahim ME, Zoorob HH. Advances in the chemistry of aminoisoxazole. Synth Commun 2013;43:2393–440.
[90] Hauser FM, Hu XD. A new route to C-aryl glycosides. Org Lett 2002;4:977–8.
[91] Tius MA, Gomezgaleno J, Gu XQ, Zaidi JH. C-Glycosylanthraquinone synthesis: total synthesis of vineomycinone B2 methyl ester. J Am Chem Soc 1991;113:5775–83.
[92] Hassner A, Rai KML. A new method for the generation of nitrile oxides and its application to the synthesis of 2-isoxazolines. Synthesis 1989;1:57–9.
[93] Albrecht HA, Blount JF, Konzelmann FM, Plati T. Synthesis of 1,2,6-thiadiazine 1,1-dioxides via isoxazolylsulfamides. J Org Chem 1979;44:4191–3.
[94] East SP, Joullie MM. Synthetic studies of 14-membered cyclopeptide alkaloids. Tetrahedron Lett 1998;39:7211–14.

CHAPTER 5

Miscellaneous use of Raney nickel for the synthesis of heterocycles

5.1 Introduction

In the comprehensive medicinal chemistry (CMC) record, above 67% of recorded compounds possess heterocyclic rings, and nonaromatic heterocyclic compounds are two times as abundant as heteroaromatics [1–3].

Heterocyclic compounds are the biggest and most diverse family of organic compounds. Among them, aromatic heterocyclic compounds represent structural motifs found in a great number of biologically active synthetic and natural compounds, agrochemicals, and medicines. Moreover, aromatic heterocyclic compounds are extensively utilized for the formation of dyes and polymeric materials of great importance. In organic synthesis, there are several reports on the use of aromatic heterocyclic compounds as intermediates. Although a variety of highly efficient approaches have been described in the past for the formation of aromatic heterocyclic compounds and their derivatives, the development of new procedures is in continuous demand. Particularly, the development of novel synthetic methodologies toward heterocyclic compounds, aiming at attaining better levels of molecular complexity and improved functional group compatibilities in a convergent and atom economical fashion from easily available starting compounds under mild reaction conditions, is one of the major research activities in synthetic organic chemistry [4–8, 9a,b].

It is easy to realize that why both the development of new approaches and the strategic deployment of known approaches for the formation of complex heterocycles compounds continue to drive the field of synthetic organic chemistry. Organic chemists have been involved in many efforts to prepare these heterocycles by developing new and efficient synthetic transformations [10].

Without doubt, the so-called Ra-Ni or Ni sponge catalyst has been the most utilized catalyst in this context. These catalysts are formed when a

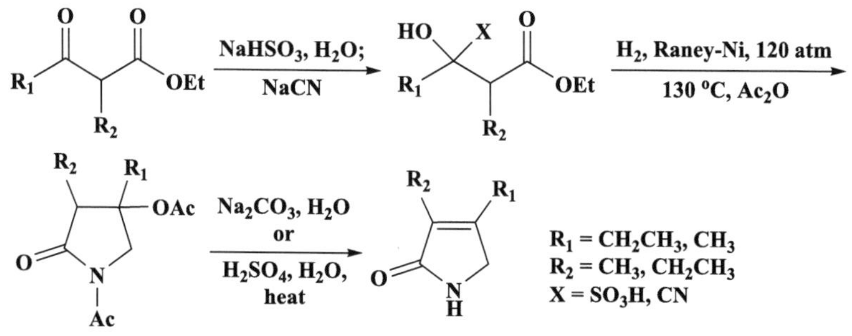

Scheme 5.1

block of Ni-Al alloy is reacted with conc. NaOH. This reaction, termed as activation, dissolves most of the Al out of the alloy. The resulting permeable structure has a large surface area, which provides high catalytic activity, obtaining different types of Ra-Ni based on the basic reaction, in all cases with a high amount of Ni (>85%), with the remaining Al helping to preserve the porous structure of the catalyst [11,12].

5.2 Synthesis of five-membered *N*-heterocycles

Improving on his cyanohydrin synthesis [13], Plieninger and Kune [14] synthesized key building blocks like 4-ethyl-3-methyl-3-pyrrolin-2-one and 3-(2'-carboxyethyl)-4-methyl-3-pyrrolin-2-one. And as revealed later, the requirement for large amounts of anhydrous hydrogen cyanide was avoided using a bisulfate adduct intermediate (Scheme 5.1) [15–17].

During his bonellin total synthesis, Montforts and Schwartz [18] made considerable development over the high temperature and pressure hydrogenation of cyanohydrins. The pyrrolinone was prepared by selective ozonolysis of dienol ether, generating acetal, and propionate chains on a double bond of formed compound. After hydrogenation (Ra-Ni, room temperature, 1 atmospheric pressure), the deprotected ester and aldehyde functions in aldehyde enabled an incorporation of nitrogen as a half-amidal, which was thermally dehydrated and isomerized to pyrrolinone (Scheme 5.2) [17].

Under the selected conditions, because of an intramolecular condensation of diamine, the only product of the catalytic reduction of 3-acetyl-4-(2-furyl)-4,5-dihydro-1*H*-pyrazole was 3-amino-2-methyl-4-(2-furyl)pyrrolodine (84% yield) (Scheme 5.3) [19].

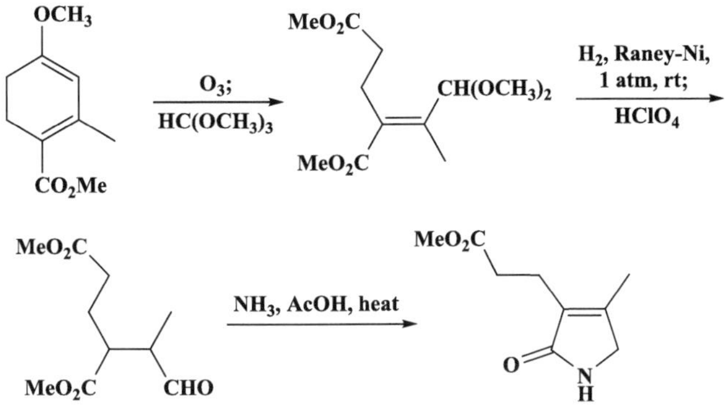

Scheme 5.2

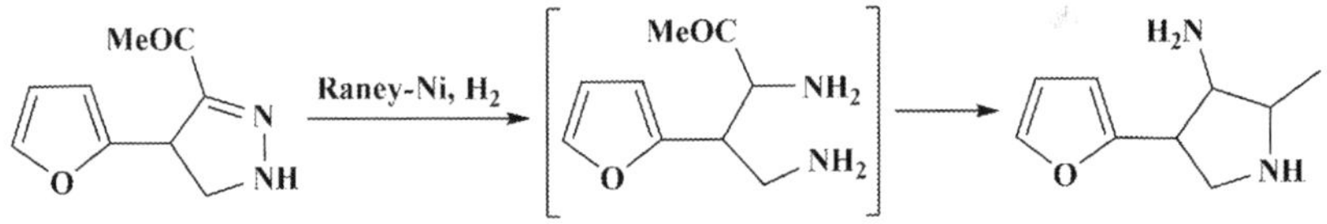

Scheme 5.3

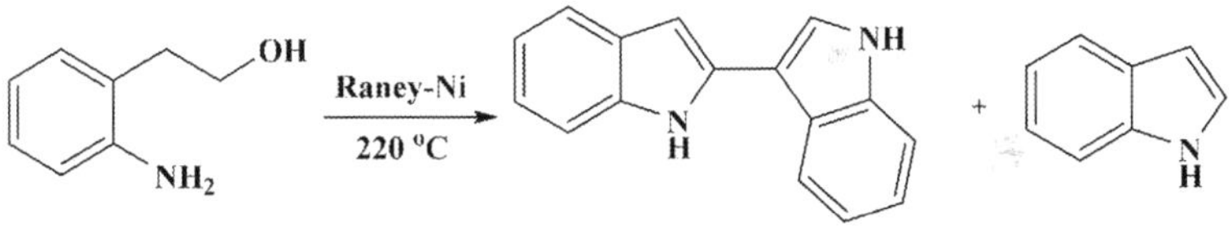

Scheme 5.4

5.3 Synthesis of five-membered *N*-polyheterocycles

It was reported that an unexpected side-product was obtained when Ra-Ni was utilized as a catalyst for this reaction (Scheme 5.4). This side-product was isolated and then purified by chromatography to afford this side-product as yellow solid in 15% yield, whose structure was assigned unambiguously as 2,3'-biindolyl by its proton nuclear magnetic resonance, ^{13}C nuclear magnetic resonance, and infrared spectra [20]. This reaction can be a beneficial technique to prepare the 2,3'-biindolyl and deserves further investigation. Moreover, it was reported that Ra-Ni maintained a stable activity after more than 20 runs in this reaction [21].

As expected, the use of degassed Ra-Ni allowed the formation of carbazole in a high yield (69%) when it was treated with phenothiazine (Scheme 5.5). It was hypothesized that oxidized sulfur may offer a better

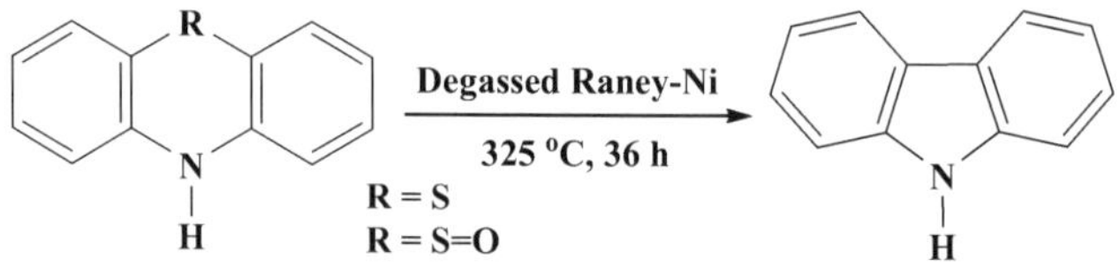

Scheme 5.5

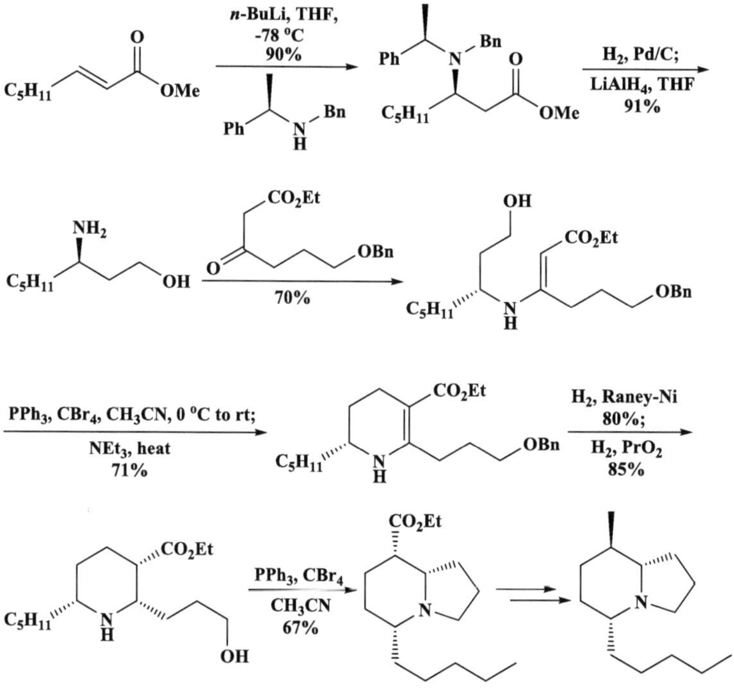

Scheme 5.6

leaving group, which allowed for a higher carbazole yield or a decline in the temperature of reaction. Although sulfur extrusion with phenothiazine-5-oxide did take place, no significant increase in yield (72%) or drop in temperature of reaction was realized [22].

5.4 Synthesis of five-membered fused *N*-heterocycles

Ma et al. [23] described a formation of (−)-indolizidine using an enantiopure β-amino ester. Remarkably, they chose to disconnect the bicyclic skeleton at the C3,C4-position (Scheme 5.6). The β-amino ester was synthesized using (*E*)-methyl octenoate. The amino alcohol was prepared

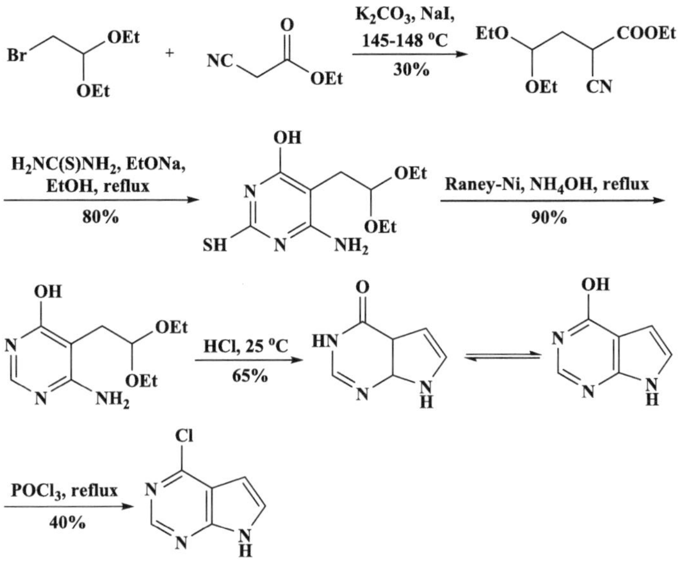

Scheme 5.7

by subsequent debenzylation and reduction of the ester utilizing LiAlH$_4$. The vinylogous urethane was synthesized by condensation of aminoalcohol with β-keto ester. The ring-closure occurred by the reaction of vinylogous urethane with CBr$_4$ and PPh$_3$ to yield the ring closed product. Hydrogenation using Ra-Ni and debenzylation provided 2,3,6-trisubstituted piperidine. Finally, the reaction between 2,3,6-trisubstituted piperidine, PPh$_3$, and CBr$_4$ provided 5,8-indolizidine, which was readily transformed into indolizidine as revealed by Michael and Gravestock [24].

The 4-chloro-7H-pyrrolo[2,3-d]pyrimidine was prepared from cheap starting compounds (Scheme 5.7) [25,26]. The treatment of bromoacetal with ethyl cyanoacetate provided ethyl diethoxyethylcyanoacetate in 30% yield after decontamination by vacuum distillation. The condensation of mononitrile with an ethanolic solution of SC(NH$_2$)$_2$ and C$_2$H$_5$ONa provided mercaptopyrimidine in 80% yield, which was further exposed to a desulfurization reaction utilizing Ra-Ni to give the pyrimidine in 90% yield. The pyrrolopyrimidone was obtained in 65% yield by spontaneous cyclization of acetal with HCl at rt. The reaction of 4-hydroxypyrrolopyrimidine with POCl$_3$ provided pure 4-chloro-7H-pyrrolo[2,3-d]pyrimidine in 40% yield after recrystallization from EtOAc.

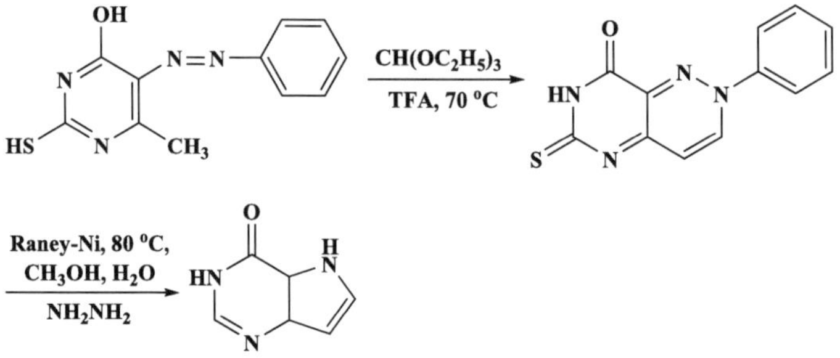

Scheme 5.8

This novel technique included the treatment of 4-hydroxy-5-phenylazo-6-methylpyrimidin-2-thione with $HC(OC_2H_5)_3$ in trifluoroacetic acid at 70°C to provide the 8-oxo-7H-2-phenylpyrimido[5,4-d]pyridazin-6-thione (Scheme 5.8) [27]. The subsequently concomitant ring contraction and elimination of 2-thione of this intermediate with Ra-Ni in CH_3OH provided desired product. In this reaction, NH_2NH_2 or sodium hydroxide could be utilized as a base catalyst [28].

Investigation of the triflaide function utilizing the model substrate, formed by a Mitsunobu reaction and subsequent hydrogenation, proved successful in the formation of fused tricycle (Scheme 5.9). It was possible to synthesize the fused indole, which cannot then be synthesized directly as previous observations showed the tendency for these allylic substrates to decompose. No reaction was observed upon application of this technique to ring-opened oxabicyclic intermediate. As there was no decomposition, it was not possible that the allylic double bond had a role in the failure of this reaction. Rather, the presence of tertiary amine poisoned the catalyst provided it was more appropriate to coordinate than the weakly coordinating triflamide [29].

The propenyl ethers of DCP have been used for the formation of hydroxy lactams [30]. The (E)- and (Z)-isomers showed different levels of selectivity when $Ti(Oi-Pr)_2Cl_2$ or MAPh was present. The (Z)-propenyl ether underwent *exo*-selective [4+2]-cycloadditions, when promoted by MAPh; on the contrary, the *endo*-selective [4+2]-cycloadditions occurred when the reactions were promoted by $Ti(Oi-Pr)_2Cl_2$. The MAPh-promoted cycloaddition of (E)-propenyl ether provided a single hydroxy lactam [(+)] formed from exclusive *exo* approach of the dienophile in the [4+2]-cycloaddition. The reactions of (E)-propenyl ether were less selective with $Ti(Oi-Pr)_2Cl_2$, giving *endo*- and *exo*-products in the ratio

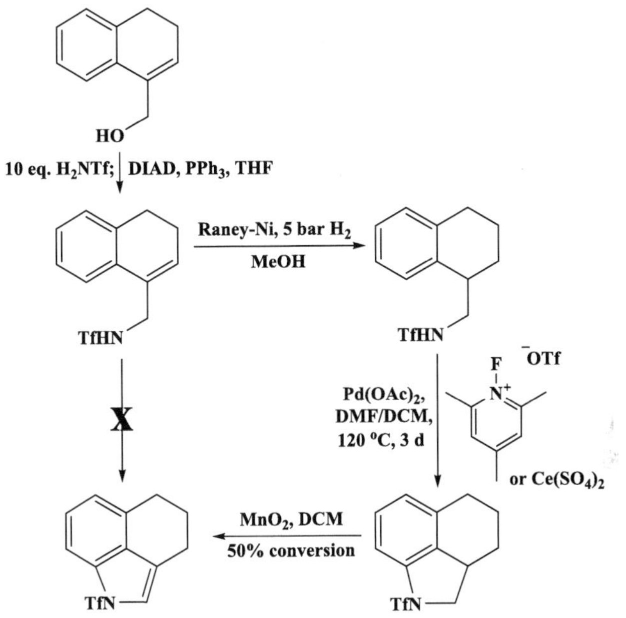

Scheme 5.9

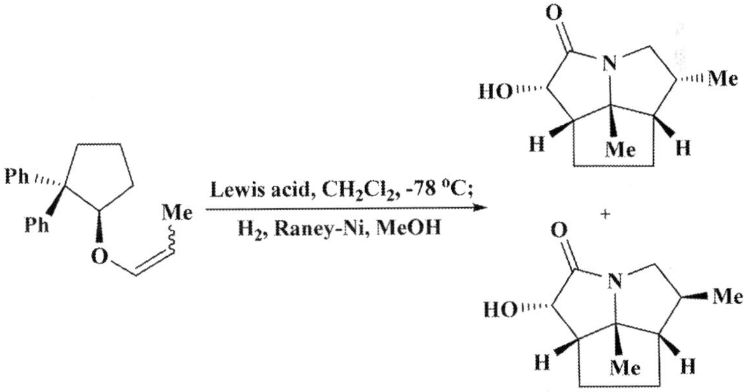

Scheme 5.10

of 1:2.3. Although the *exo*-diastereomer [(-)] was found to be highly enantiomerically enriched (96% *ee*), this erosion of *exo/endo*-selectivity can be observed as a limitation of DCP as a chiral auxiliary (Scheme 5.10).

With the oxygenated side-chain deprotected, the next stage of the synthesis was an oxidation of primary alcohol to aldehyde followed by a

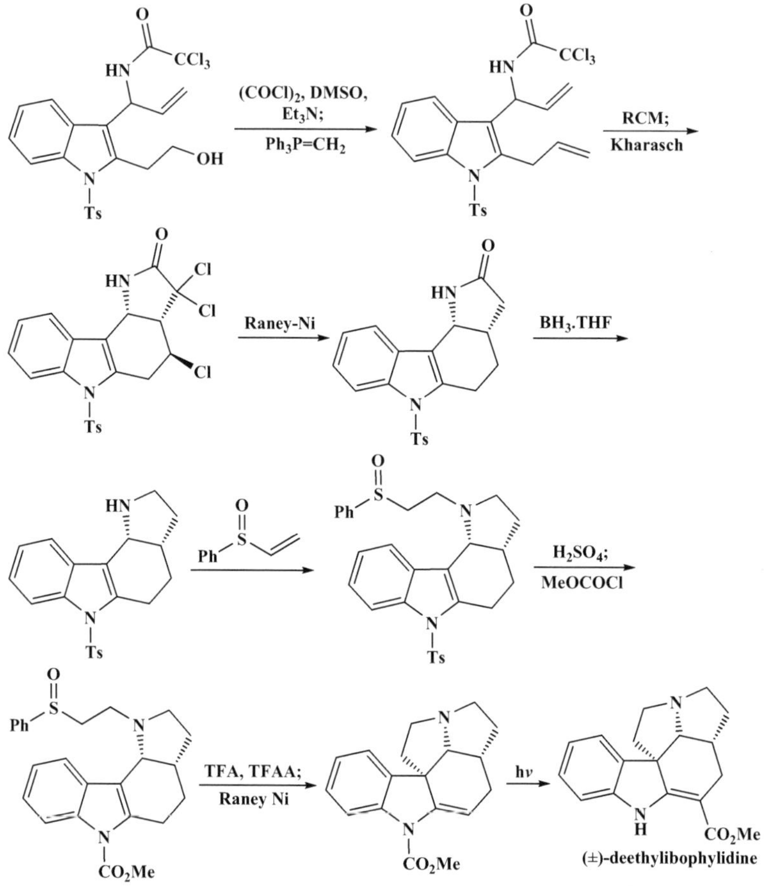

Scheme 5.11

methylenation reaction (Scheme 5.11). This will give expected diene, which will be further exposed to a tandem ring-closing metathesis/Kharasch cyclization catalyzed by Grubbs first generation catalyst. An elimination of chlorine atoms utilizing Ra-Ni™ will be followed by reduction of the lactam to provide the cyclic amine. The formal synthesis of natural product was completed by the alkylation of amine with phenyl vinyl sulfoxide followed by exchange of the tosyl group with a methyl ester [31]. To complete the total synthesis, an acid-catalyzed Pummerer rearrangement-cyclization followed by desulfurization utilizing Ra-Ni™ would afford the expected pentacyclic system. Finally, ultraviolet irradiation of this compound would deliver the desired compound.

Up to now, there have been three total syntheses of α-CPA, all starting from an appropriate 3,4-disubstituted indole. Kozikowski et al. [32] accomplished the first synthesis. They used an easily accessible N-tosyl derivative of indole-4-carboxyaldehyde and converted it to 4-(1-tosyl-1H-indol-4-yl)butan-2-one in 89% yield by a Wittig reaction followed by Pd-catalyzed hydrogenation. The ketone functionality of 4-(1-tosyl-1H-indol-4-yl)butan-2-one was protected as its ethylene ketal applying ethane-1,2-diol under acidic conditions and subsequent deprotection of N-tosyl group provided dioxane in 92% yield. Under standard Vilsmeier procedure, the aldehyde was formed by formylating 2-position of the indole, followed by use of 2-acetamido-3-methoxy-3-oxopropanoic acid, the amidoacrylate group was effectively connected and subsequent protection of the indolic nitrogen with methyl chloroformate provided amine derivative in total yield of 46% from dioxane. The deprotection of ketal followed by synthesis of thermodynamic silyl enol ether and reaction with benzenesulfenyl chloride afforded phenyl sulfide-substituted derivative of amine in 54% yield. An intramolecular Michael addition promoted by 1,8-diazabicyclo[5.4.0]undec-7-ene synthesized central carbocyclic ring in 55% yield. The Ra-Ni-induced desulfurization provided *cis*-compound almost exclusively in 72% yield. This reaction was contrathermodynamic in nature as mixing with 1,8-diazabicyclo[5.4.0]undec-7-ene led exclusively to *trans*-isomer. The closure of ring D occurred upon using thiophenol with magnesium bis-trifluoromethanesulfonate, producing α-phenylthioamide intermediate in 80% yield. A study of a multitude of organometallic reagents found that $(CH_3)_2Zn$ in $CHCl_3$ was the only reagent capable of substituting the thio group and installing the *gem*-dimethyl functionality in 73% yield. The deprotection of acetamide with triethyloxonium tetrafluoroborate (Meerwein's reagent or Meerwein's salt) in CH_2Cl_2 occurred in 79% yield followed by reaction with diketene afforded acetoacetamide in 80% yield. The iso α-CPA was formed unexpectedly in base-catalyzed cyclization of acetoacetamide to tetramic acid residue with CH_3ONa. An epimerization with TEA in $CHCl_3$ gave α-CPA:iso α-CPA in 5:2 ratio. Kozikowski and Greco completed the first formation of α-CPA in 16 steps from 1-tosyl-1H-indole-4-carbaldehyde in a total yield of 2.1% (Scheme 5.12).

The reaction was performed to obtain the cycloomologues (Scheme 5.13). The expected compound was prepared starting from diethyl 2-oxoadipate and aminopyridine in hot xylene, while diethyl 2-oxopimelate and aminopyridine afforded expected compound along with a small amount of imidazo[2,3-*b*]pyridines whose synthesis could be explained

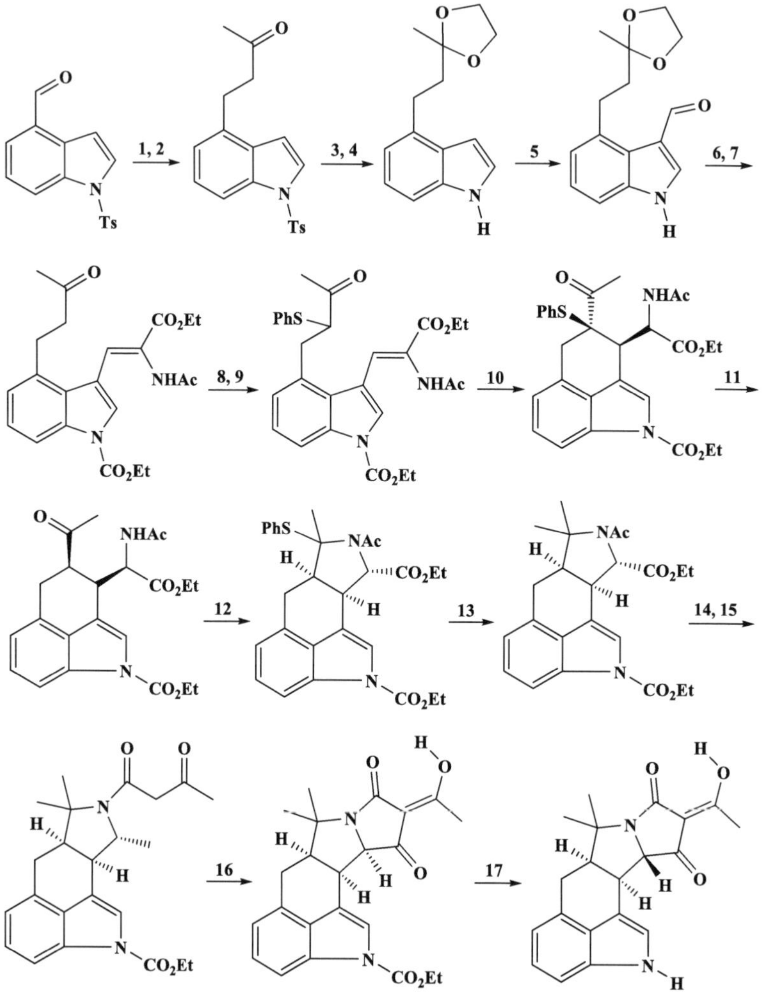

Reagents and conditions: (1) CH₃COCH=PPh₃, THF, 95%, (2) H₂, Pd/C, MeOH-THF, 94%, (3) HO(CH₂)OH, *p*-TsOH, PhH, 99%, (4) 4 N KOH, MeOH, 93%, (5) POCl₃, DMF, then NaOH, 62%, (6) ArNHCH(CO₂Et)CO₂H, Ac₂O, Py, 78%, (7) ClCO₂Et, Et₃N, 95%, (8) 10% HCl, THF, 98%, (9) MeSiI, HN(SiMe₃)₂, then PhSCl, 55%, (10) DBU, *t*-BuOH, reflux, 60%, (11) H₂, Raney-Ni, 72%, (12) PhSH, Mg(OTf)₂, 80%, (13) Me₂Zn, CH₃Cl, 73%, (14) Et₃OBF₄, Na₂CO₃, THF, CH₂Cl₂, 79%, (15) diketene, EtOH, 80%, (16) NaOMe, MeOH, PhH, (17) Et₃N, CH₃Cl, 100 °C.

Scheme 5.12

by the cyclization of an intermediate [33]. Hydrogenation with Ra-Ni under pressure followed by alkaline hydrolysis occurred without concomitant cyclization. The fusion in vacuo gave smoothly 7,7a,8,9-tetrahydro-5-methyl-5*H*-pyrido[2,3-*b*]pyrrole[1,2-*d*]diazepin-6,10-dione

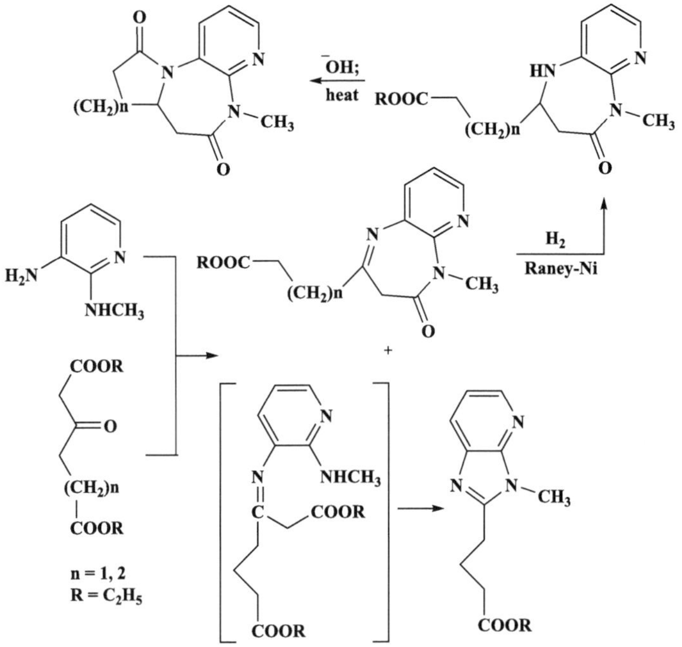

Scheme 5.13

and 7a,8,9,10-tetrahydro-7-methyl-5H-dipyrido[1,2-d:2,3-b]diazepin-6,11 (5H,7H)-dione, respectively [34-35].

5.5 Synthesis of five-membered *N,N*-heterocycles

The sulfoxides were reacted with trifluoroacetic anhydride to afford the triazapentalenes in high yields. The procedure must include the acylation of sulfoxide oxygen atom and formation of a carbocation, which attacked the N-2 atom of benzotriazole. Hydrogenation over Ra-Ni cleaved the carbon–sulfur bond and one of the nitrogen–nitrogen bonds to synthesize the *o*-substituted anilines (Scheme 5.14) [36].

The 2-adamantanyl ethanol was oxidized to aldehyde using PIDA (phenyliodine diacetate) in the presence of catalytic TEMPO ((2,2,6,6-tetramethylpiperidin-1-yl)oxyl or (2,2,6,6-tetramethylpiperidin-1-yl)oxidanyl) (Scheme 5.15). The enantioselective α-amination of aldehyde and in situ reduction gave chiral amino alcohol in 93.5% enantiomeric excess. Subsequently, the protected hydrazine was cleaved off by hydrogenation over Ra-Ni. After work-up, the crude product was

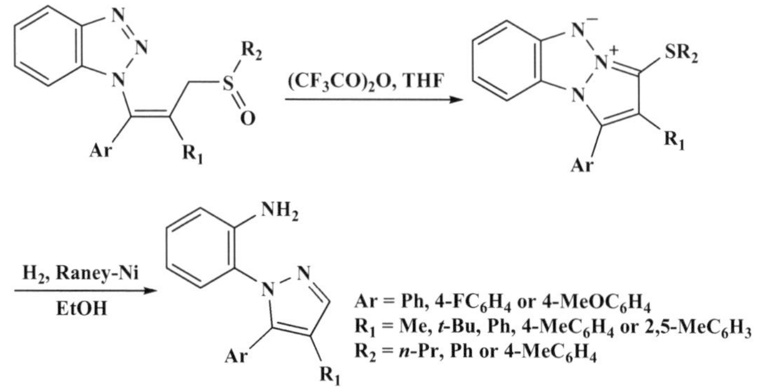

Scheme 5.14

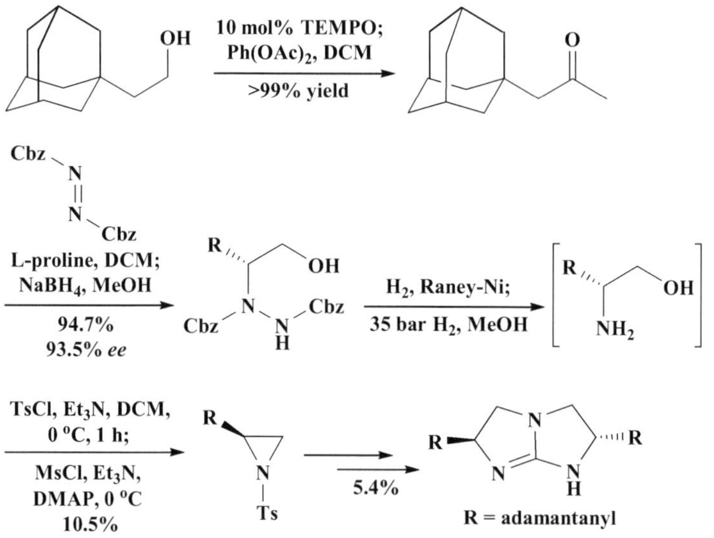

Scheme 5.15

directly subjected to cyclization reaction to give the aziridine in 10.5% yield. The low yield was due to the difficulty in ring-closure because of the presence of bulky group. The novel guanidine catalyst was obtained in 5.4% yield from aziridine [37–39].

5.6 Synthesis of five-membered N,N,N-heterocycles

Zhong et al. [40,41] applied 6-(imidazol-1-yl) directing group for the formation of 3-deaza-3-substituted purine nucleosides and needed the

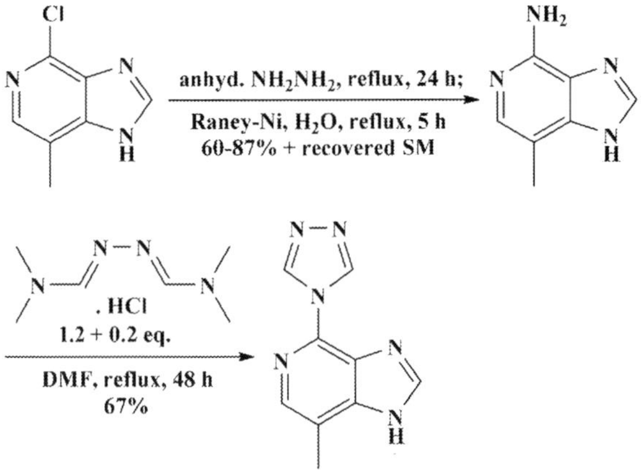

Scheme 5.16

reaction of substituted imidazole with 6-chloro-3-deazapurine (Scheme 5.16). Though the 6-chloropurines reacted with imidazoles at 65°C in dimethylformamide, 6-chloro-3-deazapurine failed to react with 2-*i*-propylimidazole even at reflux. Looking for an alternative approach of installing the imidazole directing group, the 6-chloro-3-deazapurine was reacted with anhydrous NH_2NH_2 at reflux for 24 h and then reduced with Ra-Ni in water to obtain the adenine analog in 60–87% yield. The yield was variable for this reaction, with 10–30% recovered starting compound, and the nature of the reaction made it impossible to monitor. With adenine analog in hand, the formation of imidazole ring on the exocyclic amine of adenine analog was observed for a short period of time. There are some reports of this reaction. The 6-(1,2,4-triazol-4-yl)-3-deaza-3-methylpurine was prepared by the reaction of azine reagent with adenine analog.

5.7 Synthesis of five-membered O-heterocycles

Rodriguez et al. [42] described a synthetic pathway to the core of racemic (±)- nemorensic acid by an intramolecular [5+2]-cycloaddition of substrate (Scheme 5.17). The resultant oxabicyclo[3.2.1]octane was elaborated to the core structure by Si deprotection, oxidative cleavage with lead(IV) acetate, and Jones oxidation [43]. This pathway constituted a process to access the 2,2,5,5-tetrasubstituted tetrahydrofuran cores, but it suffered from a high step count.

Noe et al. [44] reported the partial reduction of alkoxy-substituted thiophenes to enolethers, which on acidic hydrolysis afforded ketones (Scheme 5.18). On this basis they succeeded in the racemic formation

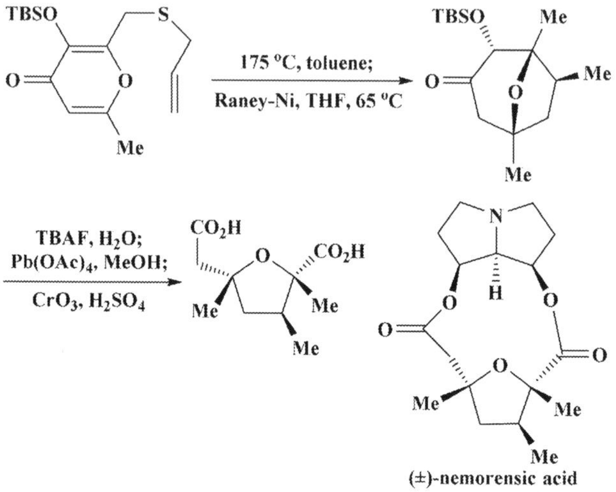

Scheme 5.17

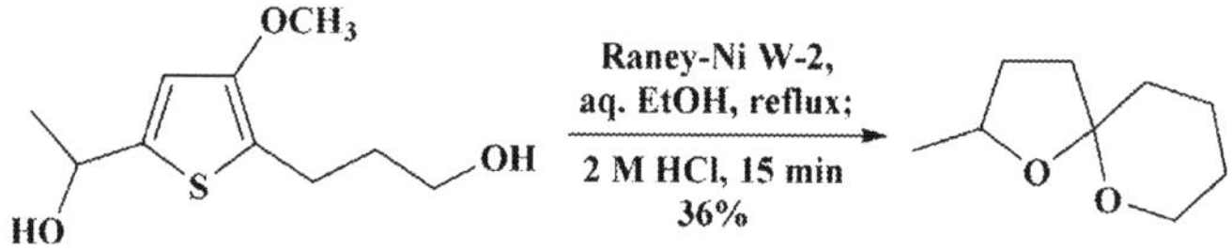

Scheme 5.18

of antiaggression pheromone of the wasp (*Paravespula vulgaris*) by reductive desulfurization of thiophene [45].

Wang and Dai [46] accomplished the total synthesis of two diastereomeric butenolide alcohols, (4S,10S,11S)- and (4S,10R,11R)-4,11-dihydroxy-10-methyldodec-2-en-1,4-olide by 1,3-dithiane bialkylation (Scheme 5.19). The *syn*-aldol subunit was constructed by a *syn*-selective aldol reaction and transformed to chiral iodide. The iodo compound along with iodinated chiral building block was utilized in the three-component linchpin reaction with 1,3-dithiane. The reductive desulfurization of coupling product with Ra-nickel led to excellent yield (93%). The selective deprotection of primary alcohol and its oxidation provided aldehyde, which was converted to alkene by a Wittig olefination. The secondary alcohol was deprotected to provide the allyl alcohol, which was further acylated to give the acrylate. The butenolide was obtained by a RCM reaction in the presence of a Grubbs II catalyst and the elimination of the tetrahydropyran ether. The linchpin coupling constitutes a flexible approach for the formation of other diastereomers [45].

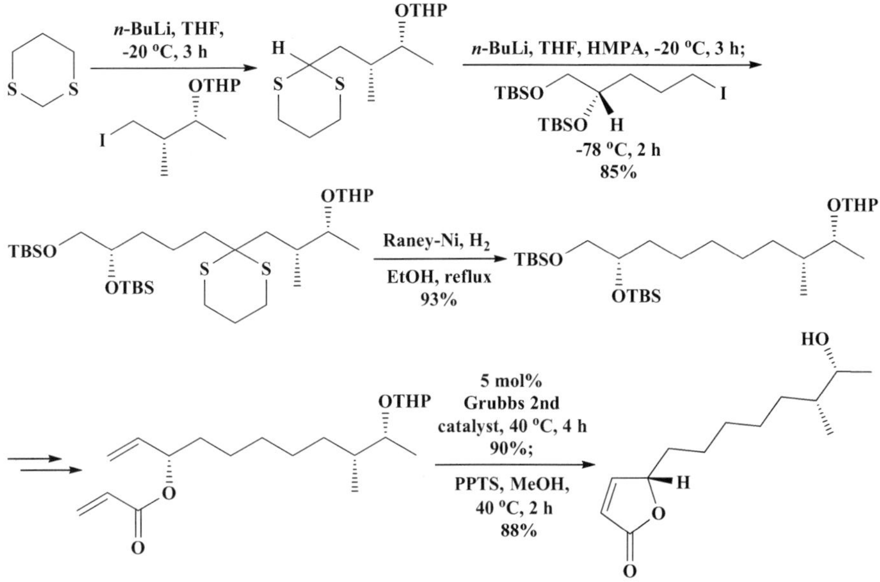

Scheme 5.19

The coupling of phosphonate and aldehyde was carried out utilizing barium hydroxide as a base at rt to provide the enone. The stereoselective reduction of enone utilizing zinc borohydride provided allyl alcohol (>98% diastereomeric excess). The reaction of allylic alcohol with Ra-Ni under hydrogen atmosphere led to the reduction of double bonds and hydrogenolysis of the C-S bond. The OH group was transformed to mesylate utilizing MsCl and triethylamine. Attempted deprotection of the acetonide utilizing aqueous 70% acetic acid to diol instead gave cyclized product with concomitant deprotection of the Mom group. The diol tetrahydrofuran derivative was treated with *t*-butyldimethylsilyl-trifluoromethanesulfonate to give the di-*t*-butyldimethylsilyl derivative, which on reaction with palladium/carbon under hydrogen pressure provided debenzylated alcohol. This alcohol was transformed to iodo derivative (Scheme 5.20). Briot et al. [47] utilized iodo derivative as an important intermediate in the formation of solamin. Thus, a formal formation of *trans*-solamin was completed.

The reaction was performed for the formation of aldehyde intermediate efficiently. The desulfurization followed by reprotection of the ketone afforded ketal in good yield. Then, an ozonolysis was carried out to give the aldehyde, which was reacted with lithiated 1,2,4-trimethoxybenzene to provide the alcohol in 59% yield (Scheme 5.21) [48].

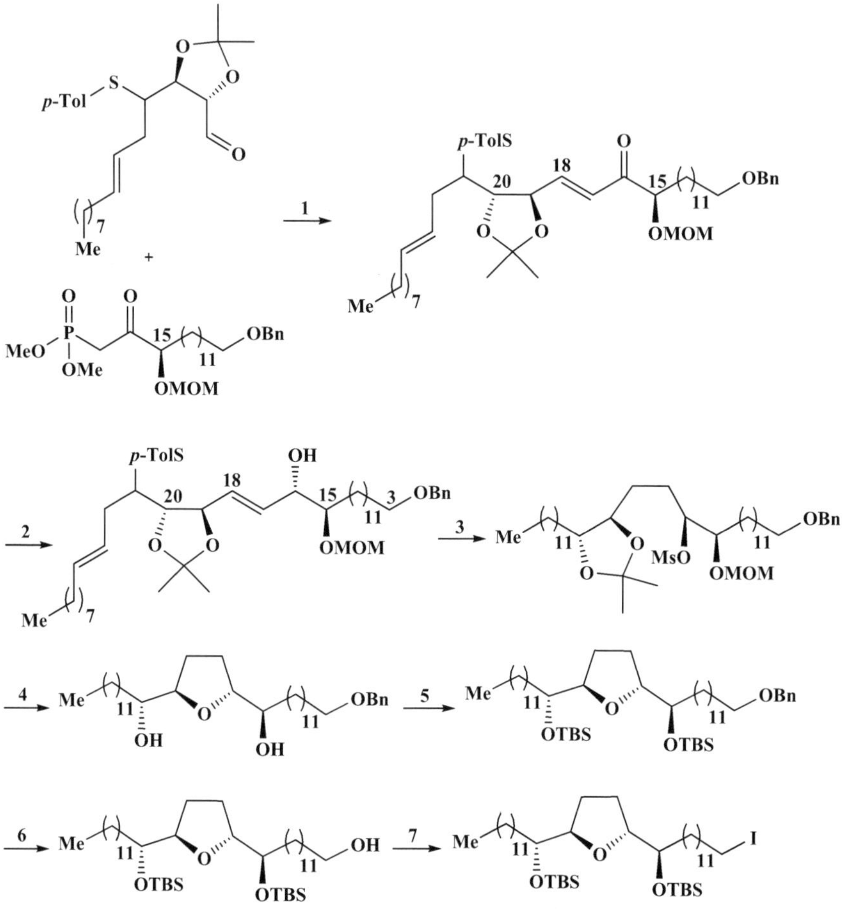

Reagents and conditions: (1) Ba(OH)$_2$, aq. THF, rt, 30 min, 70%, (2) Zn(BH$_4$)$_2$, THF, -40 °C, 2 h, 85%, (3) a) Raney-Ni/EtOH, H$_2$, rt, 2 h, 65%, b) Mesyl chloride, DCM, Et$_3$N, 0 °C, 10 min, 100%, (4) aq. 70% AcOH, 80 °C, 4 h, 65%, (5) TBSOTf, DCM, 2,6-lutidine, 0 °C, 30 min, 95%, (6) Pd/C, H$_2$, EtOAc, rt, 6 h, 95%, (7) TPP/I$_2$, imidazole, THF, 0 °C, 30 min, 71%.

Scheme 5.20

5.8 Synthesis of five-membered O,N- and S-heterocycles

List [37] and Bogevig et al. [49] independently reported a modified direct asymmetric organocatalytic amination reaction for aldehydes. Interest in this reaction stems from the versatile products that can be synthesized, like amino acids, amino alcohols, and 4-substituted oxazolidinone compounds. List [37] described a direct catalytic asymmetric amination of aldehydes

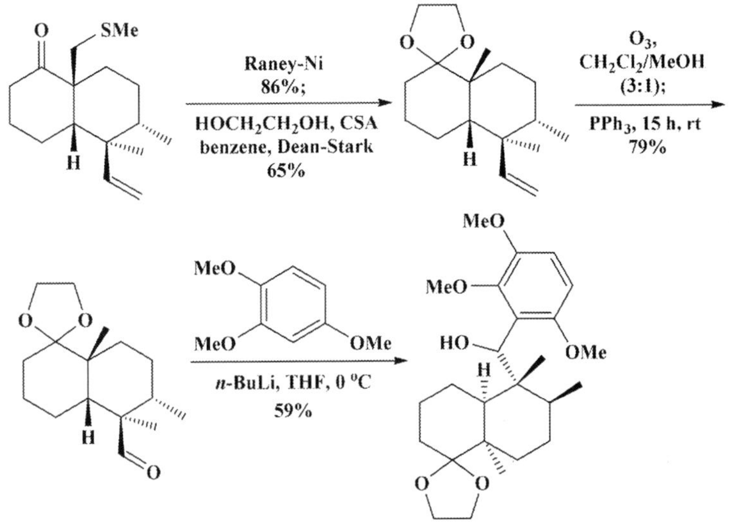

Scheme 5.21

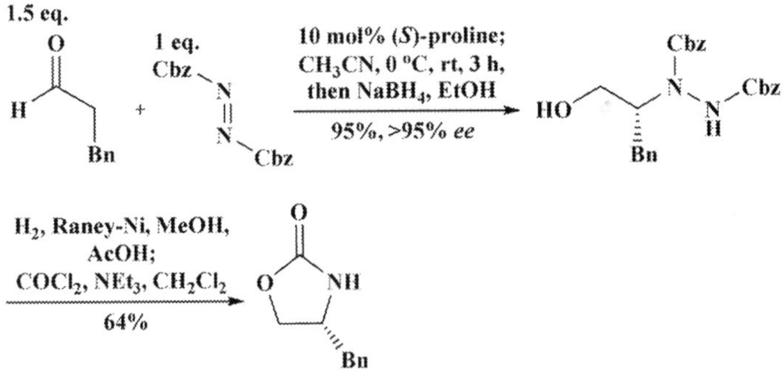

Scheme 5.22

applying (S)-proline as a catalyst and dialkylazodicarboxylate as the nitrogen electrophile with a slight excess of aldehyde in CH_3CN (Scheme 5.22). The resultant hydrazino aldehydes were reduced with sodium borohydride in situ as the product was prone to racemization. List [37] isolated the amination products as their configurationally stable crystalline 2-hydrazino alcohols with excellent enantioselectivities and high yields. The 2-hydrazino alcohols were converted to 4-substituted oxazolidinone compounds by hydrogenation over Ra-Ni and subsequent condensation with $COCl_2$.

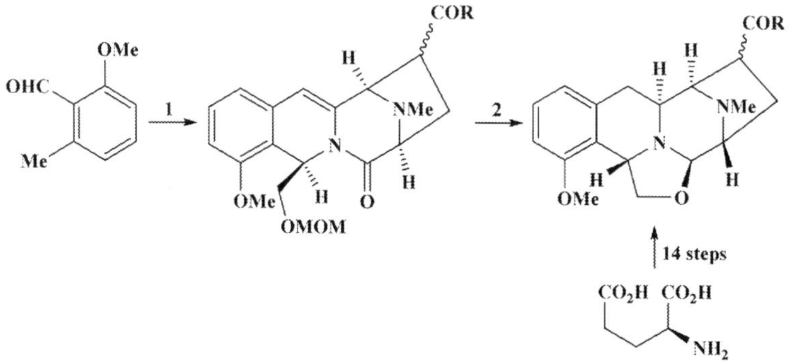

Reagents and conditions: (1) a) MeSCH₂S(O)Me, triton B, HCl, *t*-BuCOCl, Et₃N, CO₂CH(Ph)CH(Me)NLi, b) KHMDS, trisyl azide, HOAc, NaBH₄, c) H₂, Pd/C, maleic anhydride, Ac₂O, NaOAc, d) 5 N HCl, THF, MeN₃, e) hν, f) MOMCl, *i*-Pr₂NEt, NBS, hν, PPh₃, *t*-BuOK, (2) a) H₂, Raney-Ni, LiOH, b) Li/NH₃, NaCN, c) TMSCl, NaI, AgNO₃.

Scheme 5.23

Tomita et al. [50] isolated quinocarcin from *Streptomyces melanovinaceus*. It has moderate activity against Gram-positive bacteria like *Staphylococcus aureus* and *Klebsiella pneumoniae*, in addition to potent antitumor activity against a number of tumor cell lines including: MX-1 human mammary carcinoma, P388 leukemia, and St-4 gastric carcinoma. The formation of (−)-quinocarcin has been described. Garner et al. [51] reported the first asymmetric synthesis and made use of chiral auxiliaries, an intermolecular 1,3-dipolar cycloaddition and an intramolecular Wittig cyclization that gave tetracycle, and eventually (−)-quinocarcin. Unlike Garner et al. [51], Tereshima et al. [52] began their synthesis utilizing (*S*)-glutamic acid as a basis for the stereochemistry and finally afforded (−)-quinocarcin applying classical chemistry (Scheme 5.23) [53].

Owen and Peto [54] prepared cyclic oxide *trans*-3-oxabicyclo [3.3.0]octane in 67% yield although cyclic compounds containing two *trans*-fused five-membered ring are not very common. They esterified *trans*-1,2-bis-hydroxymethylcyclopentane with mesyl chloride in C₅H₅N to synthesize the *trans*-dimethanesulfonates, which on hydrolysis in aqueous alkaline medium provided cyclic oxide. In the similar study, these workers synthesized sulfur analogues of cyclic oxide. The reaction of *trans*-1,2-bis-hydroxymethylcyclopentane with TsCl afforded *trans*-ditoluene-*p*-sulfonate, which on further treatment with potassium thiolacetate and thiolacetic acid in C₂H₅OH provided *trans*-1,2-di(acetylthiomethyl)cyclopentane. This bis-thiolacetate was hydrolyzed to dithiol and the resultant solution was treated with Ra-Ni to provide the *trans*-3-thiabicyclo[3.3.0]octane (Scheme 5.24).

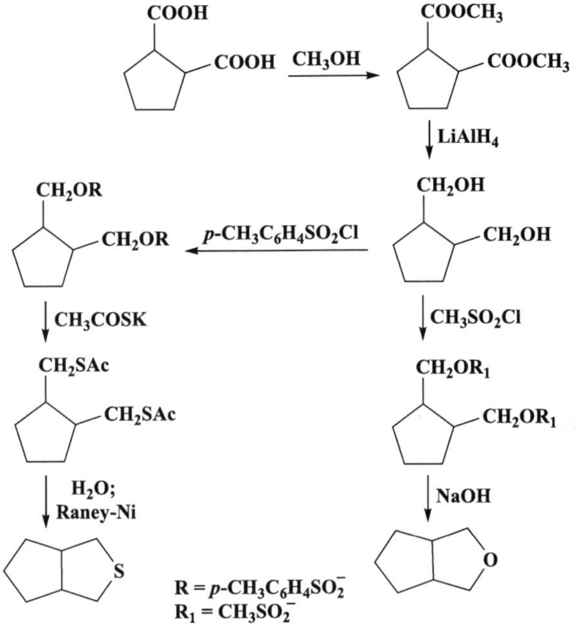

Scheme 5.24

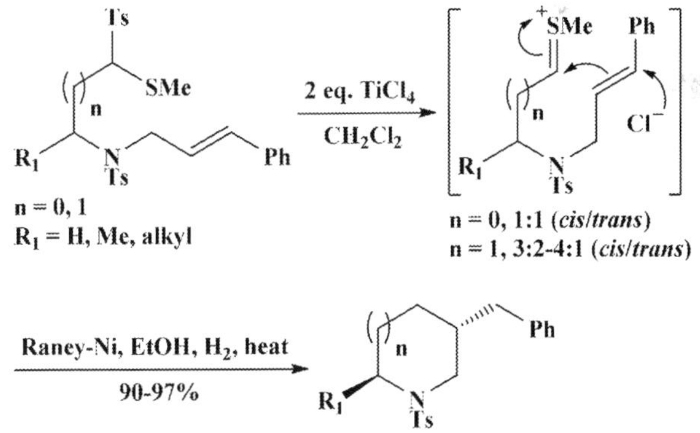

Scheme 5.25

5.9 Synthesis of six-membered *N*-heterocycles

Craig et al. [55] reported the formation of piperidines and pyrrolidines by thionium ion–assisted cyclization of suitably substituted dithioacetal *S,S*-dioxides (Scheme 5.25). The reaction of dithioacetal *S,S*-dioxides with TiCl$_4$ at rt led to thionium ion (by removal of toluene-*p*-sulfinate), which further cyclized in a nonselective manner to provide a complex mixture of

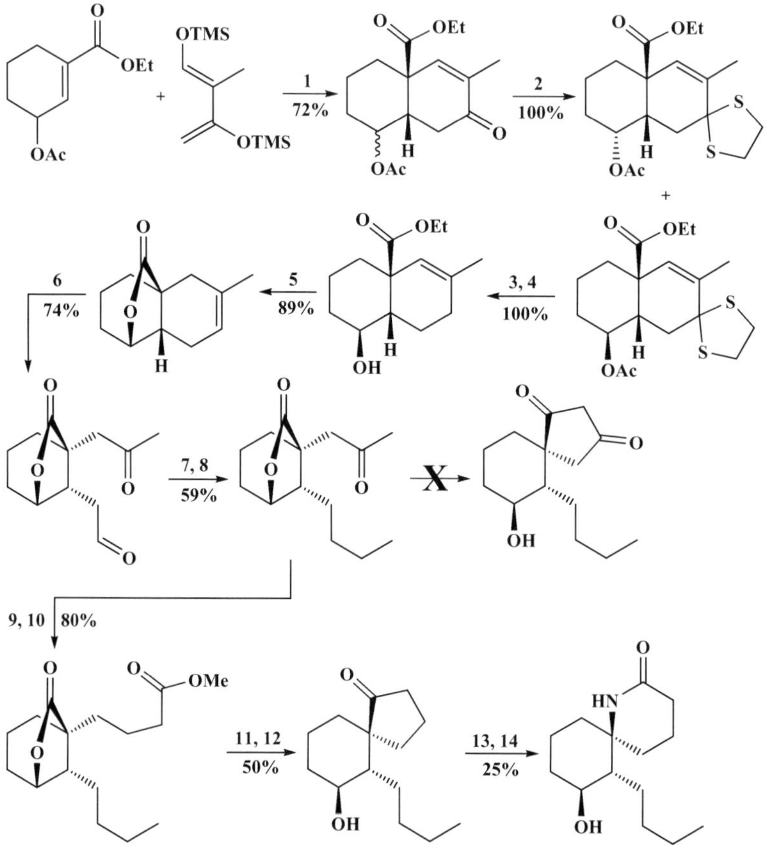

Ibuka's 1982 (±)-pHTX formal synthesis. Reagents and conditions: (1) 190 °C; HCl, H$_2$O, (2) HS(CH$_2$)SH, (3) Raney-Ni, (4) KOH, H$_2$O, (5) *p*-TsOH, toluene, reflux, (6) OsO$_4$, NaIO$_4$, THF, -50 °C, (7) ethyltriphenylphosphonium bromide, NaH, -50 °C, (8) H$_2$, PtO$_2$, (9) Whitlock methoxycarbonylation, (10) NaBH$_4$, MeOH, -40 °C; MsCl, Py; DBU, Et$_3$N; H$_2$, PtO$_2$, (11) KOH, THF, (12) DABCO, xylene, reflux, (13) NH$_2$OH-HCl, NaHCO$_3$, MeOH, (14) TsCl.

Scheme 5.26

isomeric piperidines. The reduction with Ra-Ni simplified the mixture to somewhat providing pyrrolidine (n = 0) or piperidine (n = 1) as mixtures of isomers [56].

Ibuka et al. [57] reported a synthetic pathway for the synthesis of lactam, which started with Diels-Alder reaction between ene and diene to provide the bicyclic enone as an inseparable mixture of *O*-acyl epimers (Scheme 5.26). This mixture was then transformed into its thioketals providing a separable 85:15 mixture in favor of the desired epimer. The hydroxy ester was prepared by desulfurization followed by acetal hydrolysis, which on thermodynamic double bond migration easily transformed to lactam in 89% yield. The keto aldehyde was synthesized by an oxidative cleavage of the alkene, which underwent a smooth chemoselective Wittig homologation to give the keto lactone on hydrogenation. The planned intramolecular Claisen addition disappointingly failed to provide the diketone, and they were forced to examine the alternate pathways. It was found that methyl ester was obtained by methoxycarbonylation of kinetic enolate of keto lactone followed by reduction of resulting ketone. The Dieckmann condensation followed by elimination of methoxycarbonyl group utilizing 1,4-diazabicyclo[2.2.2]octane (DABCO) in refluxing xylene provided spirocyclic hydroxy ketone. This was further transformed to lactam utilizing earlier developed conditions for the key Beckmann rearrangement [58,59].

This methodology included the formation of a piperidine ring with a quaternary center placed α to the nitrogen. The conventional C-C bond forming reactions are then used to complete the spirocyclic ring system. Various methods including this approach have been based on the ring-contraction of cyclic enamino aldehydes to synthesize the quaternary piperidine ring rapidly. The first group to use this method was that of Duhamel et al. [60], in a formal synthesis of pHTX (Scheme 5.27). The aldehyde was prepared in excellent yield by formylation of thiolactam with Bredereck's reagent, followed by hydrolysis. The reaction of this aldehyde with methyl trifluoromethanesulfonate followed by Et_3N-assisted deprotonation synthesized methylthioaldehyde. The methylthioaldehyde underwent a Ra-Ni-assisted desulfurization to prepare the enamino aldehyde. The desired ring-contraction proceeded in near-quantitative yield to give the acetal when enamino aldehyde was treated with bromine and Et_3N in CH_3OH. The condensation with CH_3COCH_3 provided unsaturated keto acetal, which underwent a Pd-catalyzed hydrogenation to synthesize an intermediate. The acetal hydrolysis then promoted the spontaneous cyclization to generate the Pearson intermediate [59].

A popular pathway to chiral heterocyclic building blocks was described by Bringmann et al. [61] (Scheme 5.28). It included an addition of nitrogen moiety via Henry reaction. The subsequent reduction of nitrostyrene with Fe powder provided ketone, which on further treatment with (S)-methylbenzylamine provided chiral imine. The stereoselective hydrogenation with Ra-Ni, followed by deprotection, liberated the chiral amine, which was easily transformed to dihydroisoquinoline by standard proces.

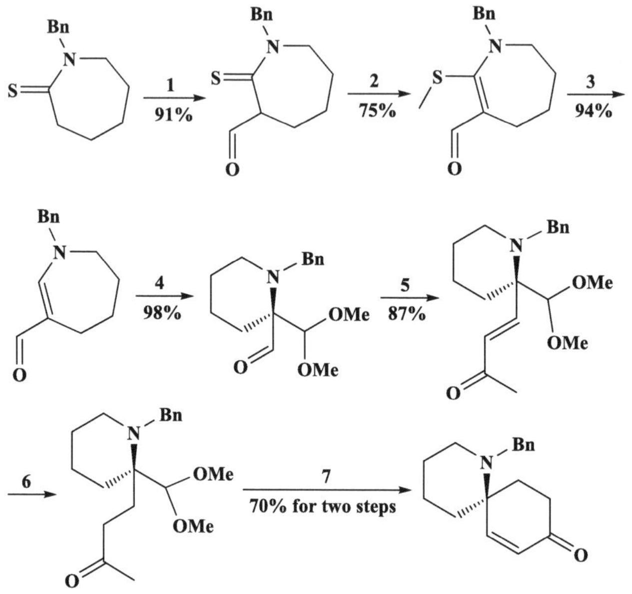

Duhamel's 1986 (±)-pHTX formal synthesis. Reagents and conditions: (1) *t*-BuOCH(NMe$_2$)$_2$, 140 °C, (2) MeOSO$_2$F, CH$_2$Cl$_2$, Et$_3$N, (3) Raney-Ni, acetone, (4) Br$_2$, Et$_2$O, -70 °C; MeOH, Et$_3$N, -70 °C, (5) *t*-BuOK, acetone, THF, -10 °C, (6) H$_2$, Pd/C, KOH, EtOH, EtOAc, (7) 3 M HCl, reflux.

Scheme 5.27

The stereoselective reduction with lithium aluminium hydride.trimethylaluminium provided *trans*-tetrahydroisoquinoline in 85% yield and 92% diastereomeric excess (*de*). Although the reaction conditions were not exceedingly tolerant of moiety, and the imines are often reported as unstable brown oils, chromatography can be avoided by the purification of amines as their hydrobromide or hydrochloride salts.

This synthesis (Scheme 5.29) was started by preparing tetrahydroisoquinolinol from 3,4-dimethoxybenzaldehyde and the aminoacetaldehyde acetal via a Pomeranz-Fritsch [62,63] reaction. The aminoalcohol was formed as a mixture of diastereomers by nitrozation of tetrahydroisoquinolinol with NaNO$_2$, followed by lithiation and alkylation along with 3,4-methylenedioxybenzyl bromide and subsequent in situ denitrozation with lithium aluminium hydride. Then, the amine was protected and the ketone, formed by an oxidation of alcohol under phase-transfer conditions, underwent a 1,2-addition with lithio-nitroso-pyrrolidine. The elimination of nitroso group with Ra-Ni, followed by reduction of *N*-(2)-benzoyl to

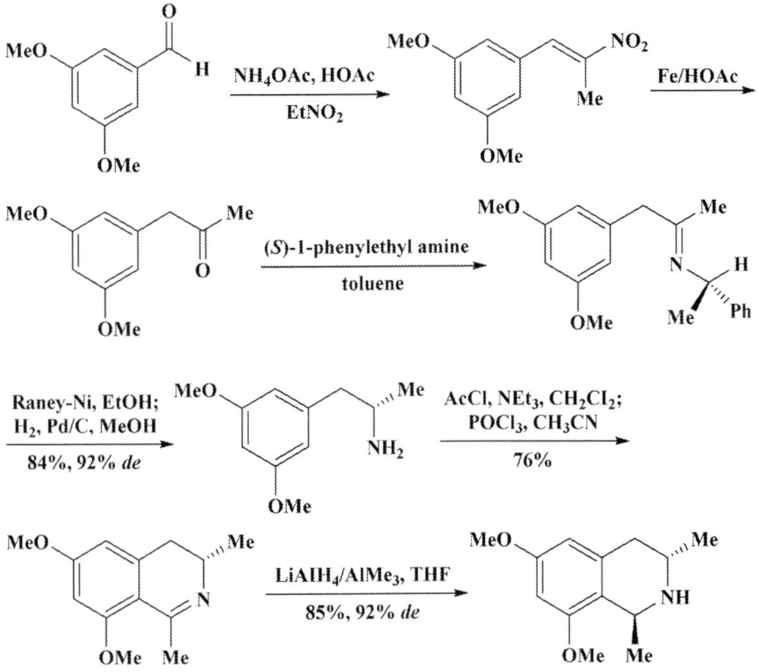

Scheme 5.28

benzyl group, and formylation of the nitrogen of pyrrolidine occurred. Finally, the formed compound was subjected to dehydration and the cleavage of N-benzyl group with palladium/carbon followed by lithium aluminium hydride reduction of the formamide to achieve the first synthesis of (±)-desired compound.

Various polyhydroxylated quinolizidines [64], frequently designed as azasugars, are powerful glycosidase inhibitors and have potential therapeutic uses. The 7-oxa-1-azabicyclo[2.2.1]heptanes, synthesized from 3-O-benzyl-1,2-O-i-propylidene-1,5-pentadialdo-D-xylofuranose by reacting with N-(1,1-dimethylbut-3-enyl)-hydroxylamine followed by an intramolecular 1,3-dipolar cycloaddition, were easily transformed to polyhydroxylated quinolizidine by elimination of i-propylidene group and hydrogenolysis of nitrogen–oxygen bond using Ra-Ni accompanied by an intramolecular reductive amination (Scheme 5.30) [65].

The azabicyclic skeleton of quinolizidine is an important structural subunit present in several alkaloids. The wide range of biological activities found in this family of natural products make them ideal targets for overall

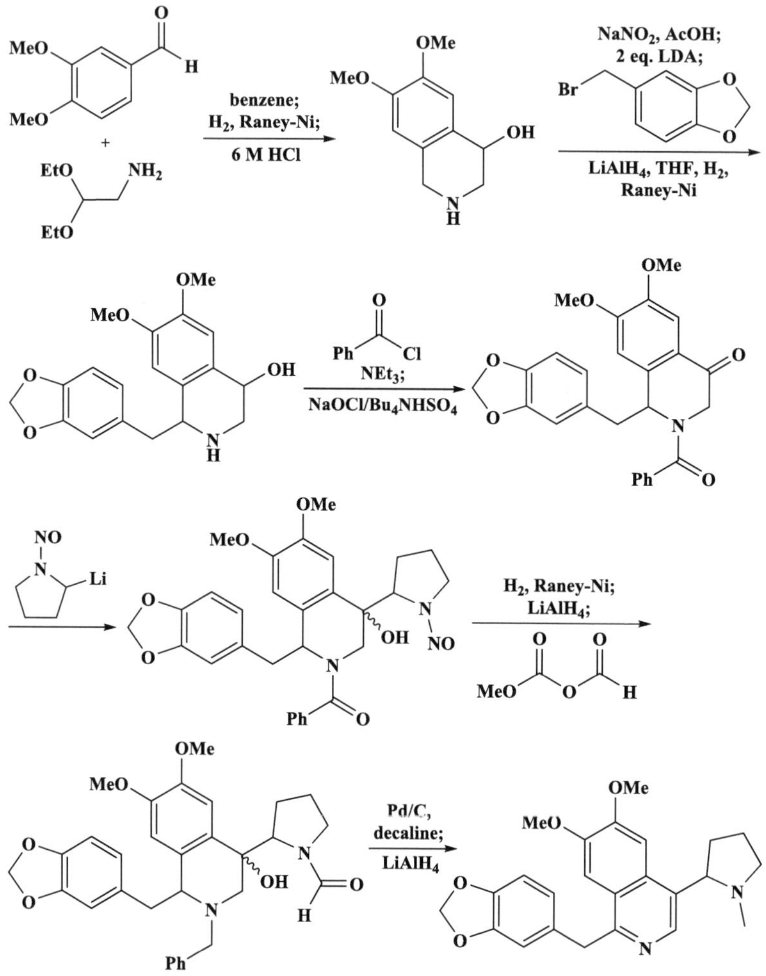

Scheme 5.29

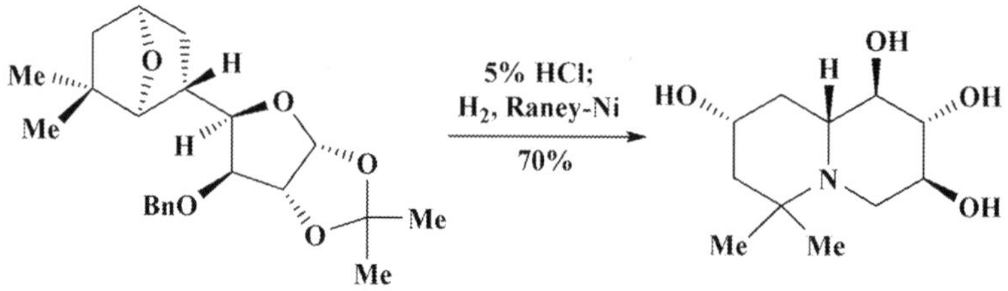

Scheme 5.30

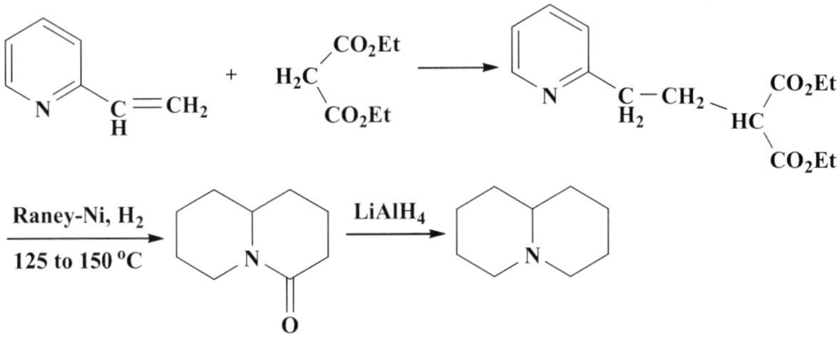

Scheme 5.31

formation. The quinolizidine was synthesized by Boekelheide and Rothchild [66]. The 1,1-dicarbethoxy-3-(2-pyridyl)-propane was synthesized by Michael addition of diethyl malonate to vinyl pyridine. Bohlman et al. [67] reported a process for the formation of 4-quinolizidone using Ra-Ni hydrogenation of 1,1-dicarbethoxy-3-(2-pyridyl)-propane in dioxane solution at 125 to 150°C under H_2 pressure. The lactam was then transformed to quinolizidine in high yield by the action of LiAlH$_4$ (Scheme 5.31).

Veerasamy et al. [68,69] reported this methodology for the formation of cermizine D (Scheme 5.32). The piperadine aldehyde was used as a starting compound, serving as the source of two of the three rings of cermizine D [70]. A number of pathways have been developed for the transformation of aldehyde to sulfone, the shortest being two steps and the longest is of eight steps. The sulfone was coupled with another unit of aldehyde to afford the sulfone alcohol. The reaction of sulfone with Ra-Ni followed by treatment with hydrochloric acid synthesized amino alcohol. The cermizine D was synthesized by Appel reaction of amino alcohol.

In this 16-step methodology (Scheme 5.33), the commercially accessible amine was utilized as the starting compound. The heterocyclic compound was generated after Boc-protection and cross-metathesis, an intramolecular Michael addition. The homologation by Wittig olefination and enol ether cleavage afforded aldehyde. The phenyl sulfide-substituted compound was prepared by Pinnick oxidation [71] of aldehyde, followed by reaction with Evans's oxazolidinone [72] and diastereoselective alkylation. The reduction of phenyl sulfide-substituted compound with LiBH$_4$ followed by sulfide oxidation and deoxygenation delivered sulfone. The hydroxysulfone as a diastereomeric mixture was formed by deprotonation of sulfone and reaction with earlier synthesized aldehyde. An oxidation of undesired

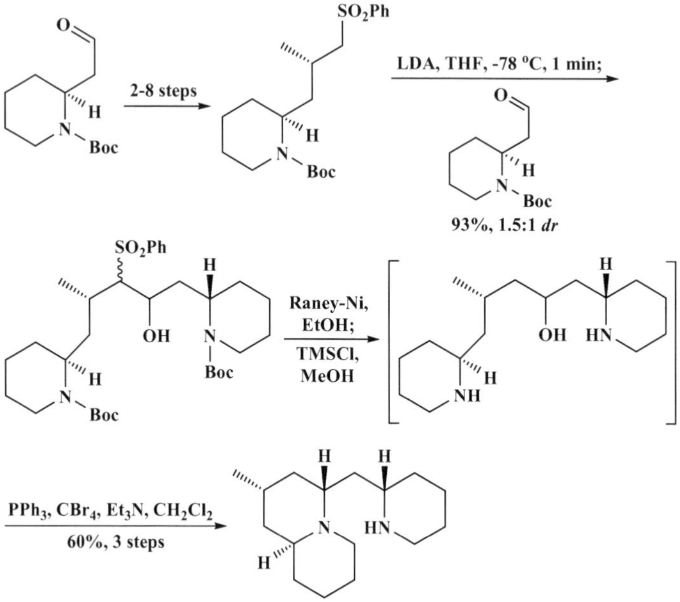

Scheme 5.32

diastereomer and subsequent reduction afforded a near equal diastereomeric mixture. The (+)-target compound was constructed by desulfurization of hydroxy sulfone with Ra-Ni followed by deprotection and cyclization.

Strelchyonok et al. [73] described the formation of 8-azasteroid by cyclocondensation of 6-methoxy-3,4-dihydroisoquinoline wi 1-acetyl-cyclopentene to provide the stereoisomeric 8-methoxy-1,2,3,3α,5,6,10β, 11,12,12α-decahydrocyclopenta[5,6]pyrido[2,1-a]isoquinolin-12-one, which was further transformed to ethanedithiol cyclic ketals and further reduction with Ra-Ni gave 8-azasteroid (Scheme 5.34).

The main challenge in designing an efficient synthesis of *Corynanthe* alkaloids is to control the relative stereochemistry of the C-3, C-15, and C-20 stereocenters around the D-ring. Brown et al. [74] planned accomplishing this task by setting the relative stereochemistry around a five-membered ring in which eclipsing 1,2-interactions would allow approximately complete diastereoselectivity. As shown in Scheme 5.35, this would allow access to synthetic dihydrosecologanine aglucone analogue. Given the success in transforming dihydrosecologanine into several indole alkaloids (*vide supra*) together with hirsutine, Brown et al. [74] anticipated an efficient entry to *Corynanthe* class of alkaloids via a synthetic analogue of dihydrosecologanine. The cyclopentene dimer was formed in 50% yield by vinylogous

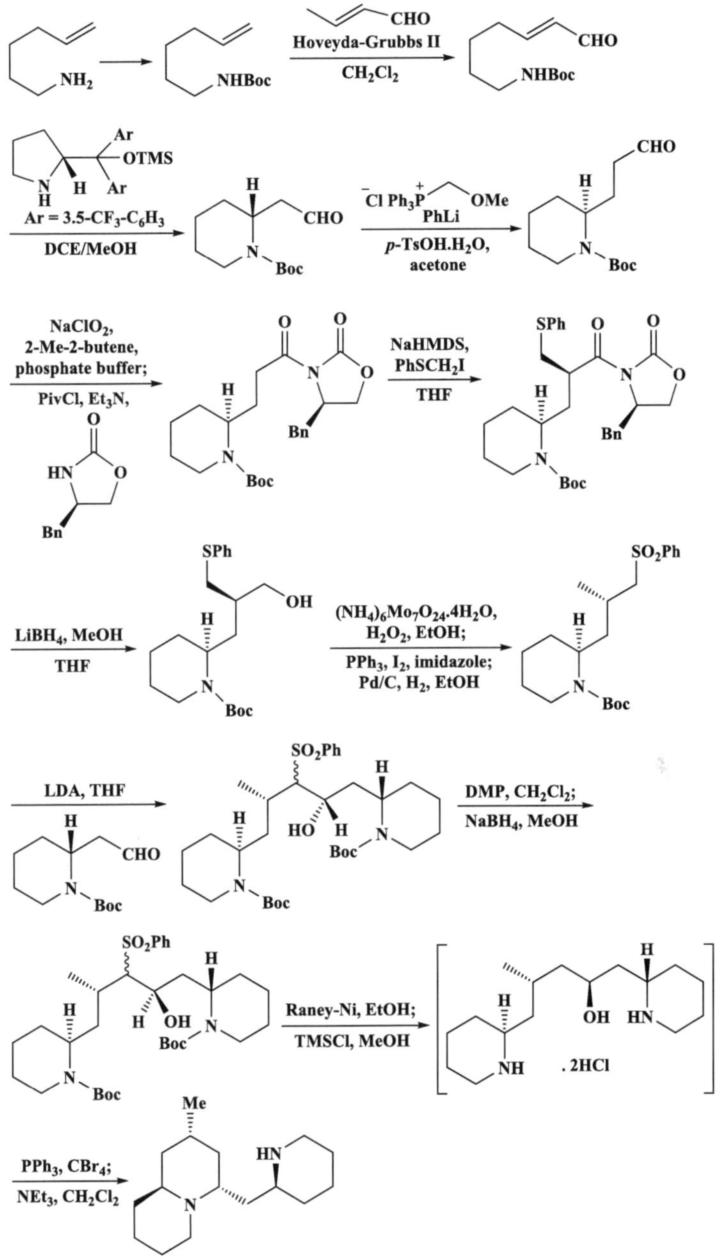
NH2
NHBoc
CHO
Hoveyda-Grubbs II
CH2Cl2
CHO
NHBoc
Ar
OTMS
H Ar
Ar = 3.5-CF3-C6H3
DCE/MeOH
CHO
Boc
Cl Ph3P OMe
PhLi
p-TsOH.H2O,
acetone
CHO
Boc
NaClO2,
2-Me-2-butene,
phosphate buffer;
PivCl, Et3N,
Bn
NaHMDS,
PhSCH2I
THF
SPh
LiBH4, MeOH
THF
OH
(NH4)6Mo7O24.4H2O,
H2O2, EtOH;
PPh3, I2, imidazole;
Pd/C, H2, EtOH
SO2Ph
LDA, THF
DMP, CH2Cl2;
NaBH4, MeOH
Raney-Ni, EtOH;
TMSCl, MeOH
. 2HCl
PPh3, CBr4;
NEt3, CH2Cl2
Me

Scheme 5.33

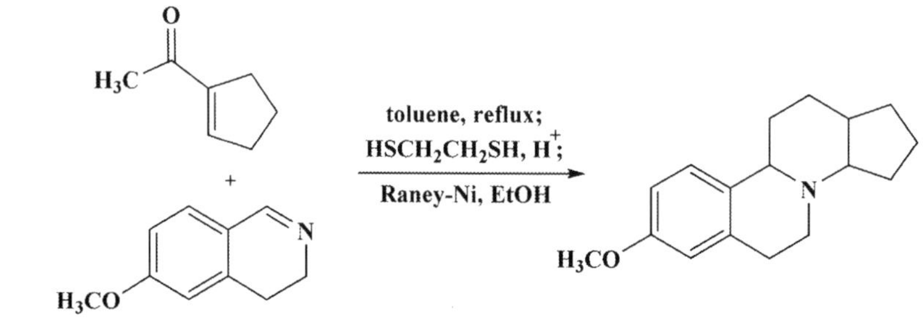

186   Raney nickel-assisted synthesis of heterocycles

Scheme 5.34

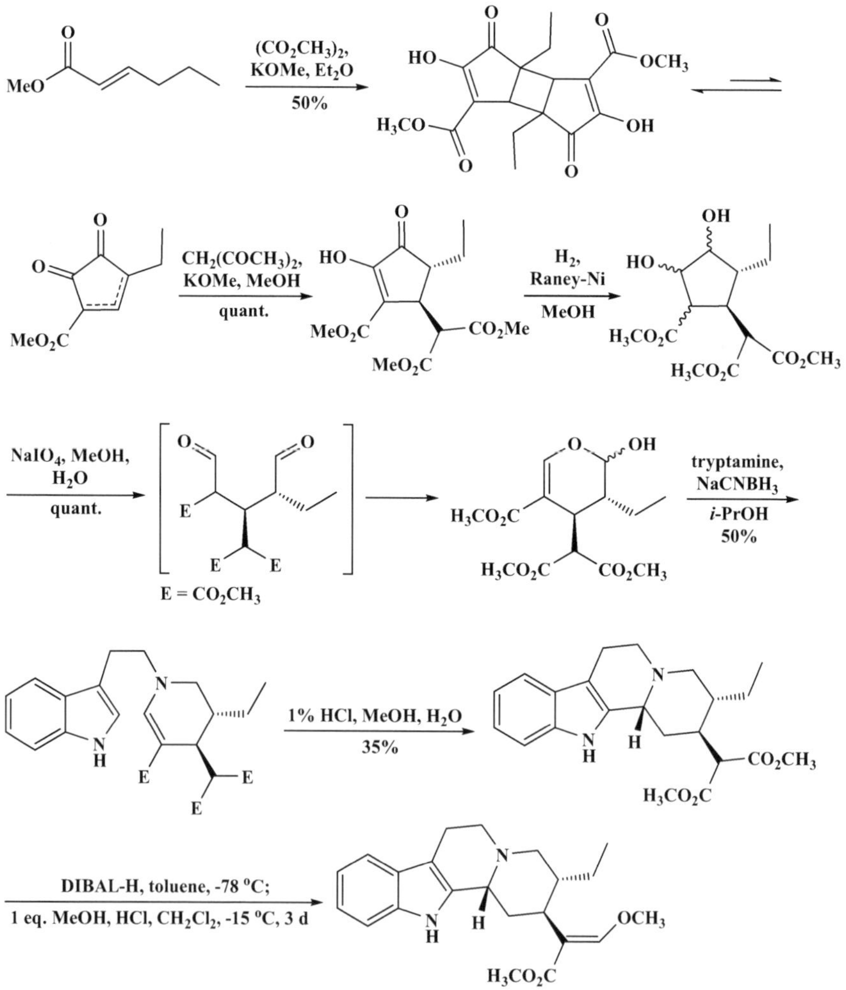

Scheme 5.35

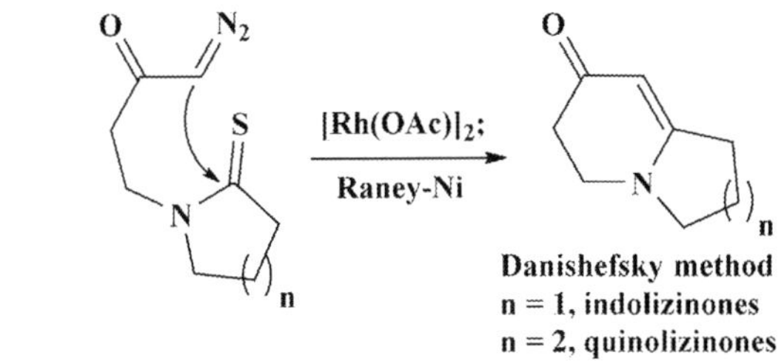

Scheme 5.36

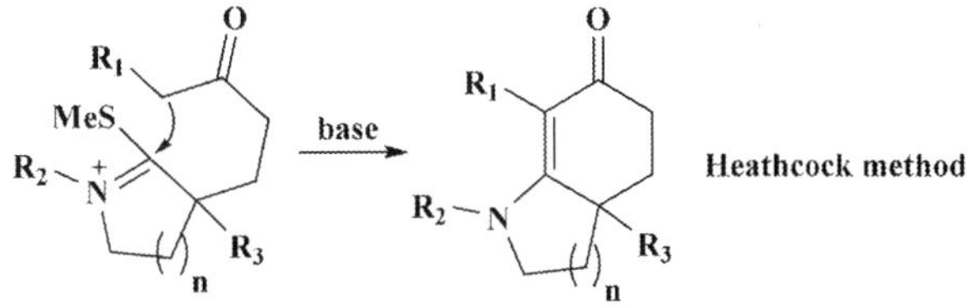

Danishefsky method
n = 1, indolizinones
n = 2, quinolizinones

Scheme 5.37

Heathcock method

Claisen condensation of hex-2-enoate with dimethyloxalate. This material exists in an equilibrium with cyclopentenones, which underwent a highly diastereoselective Michael addition when reacted with Et$_3$N and dimethylmalonate to deliver a product in quantitative yield as a single stereoisomer. The hydrogenation over Ra-Ni gave diol, which cleaved when treated with NaIO$_4$ to provide the secologanine analogue. The reductive amination with tryptamine provided a product in 50% yield, which led to a one-pot ester hydrolysis/decarboxylation protocol followed by an intramolecular stereoselective Pictet-Spengler reaction to give the indoloquinolizidine in 35% yield. The transformation of indoloquinolizidine to hirsutine was performed utilizing the method, which was previously reported by Wenkert et al. [75]. While the synthesis was rather concise, it was not easily adapted to an enantioselective synthesis. This issue was ultimately addressed by Brown et al. [74] around 6 years later, resulting in the first enantioselective synthesis of hirsutine (*vide infra*).

Kim et al. [76] prepared bicyclic indolizinones and quinolizinones by a rhodium(II)-catalyzed reaction of thiolactams and α-diazoketones via ring-closure reaction (Scheme 5.36) [77]. However, the use of CH$_2$N$_2$ for the formation of α-diazoketone could restrict the scope of this approach to small scale synthesis. To solve this problem, a new strategy towards the development of aza-Robinson annulations-type reaction on a large scale has been described. This alternative included the transformation of thiolactam into an iminium ion, which further underwent an intramolecular attack by internal enolate on the side-chain (Scheme 5.37). This approach was first

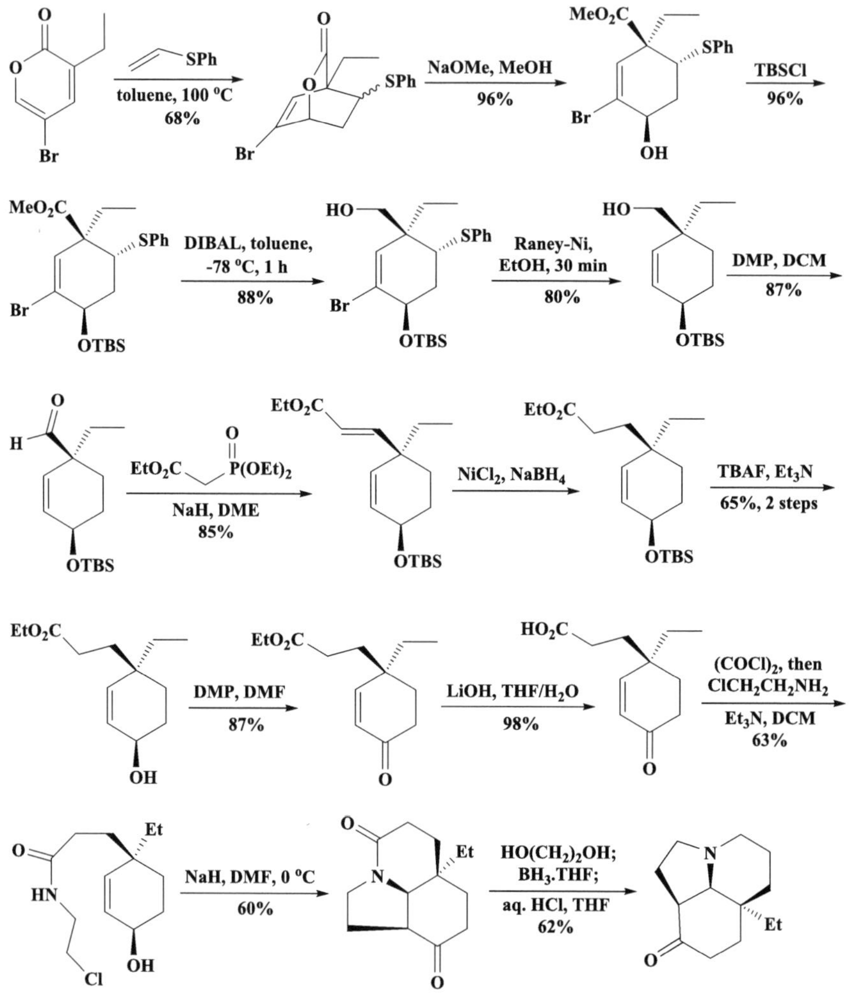

Scheme 5.38

reported by Heathcock et al. [78] and then used by others [79].

The synthesis started with a palladium-catalyzed incorporation of ethyl group on the C3-position of 3,5-dibromo-2-pyrone [80]. The use of triethylaluminium-dimethylaminoethanol complex provided the best results in terms of selectivity and yield [81]. Upon securing multigram quantity of the 2-pyrone, the Diels-Alder reaction was carried out with phenyl vinyl sulfide which provided bicyclolactone as a mixture of *exo/endo*-isomers (2:3, 68% combined yield, Scheme 5.38). Much similar to the cycloaddition reactions of other C3-substituted 2-pyrones, the ethyl group at C3-position is presumed to destabilize otherwise favored *endo*-transition state, resulting in moderate *exo/endo*-selectivity [82]. Although both isomers were tactically equivalent (phenylthio moiety was eliminated later in the sequence), the *exo*- and *endo*-cycloadducts were separated and carried independently through the reaction sequence for the handiness in the spectroscopic characterizations (only the reaction with *endo*-isomer is shown) [83]. The lactone opening of bicyclolactone and protection of OH group as a TBS ether occurred in good yield. The methyl ester was then reduced to provide the alcohol, prior to the reductive elimination of both phenylthio and vinyl bromide groups with Ra-Ni to obtain the cyclohexene. The ethyl enoate was synthesized by a Dess-Martin oxidation to aldehyde followed by a Wittig olefination with phosphonate. The selective conjugate reduction of enoate double bond provided ester together with inseparable over-reduced product in the ratio of 6:1. The removal of TBS protecting group allowed the separation and provided pure alcohol in 65% yield over 2 steps. The acid was formed in good overall yield by an oxidation of allylic alcohol to ketone and hydrolysis of the methyl ester. The synthesis of acid chloride followed by coupling reaction with chloroethylamine gave amide in 63% yield. The tricyclic ketone was obtained in 60% yield when enone was reacted with sodium hydride in dimethylformamide by tandem conjugate addition-alkylation cascade. Its amide group was further reduced to synthesize the ketone by following the known three-step reaction sequence involving ketalization, amide reduction, and ketal unmasking [84].

Oppolzer et al. [85] described an incorporation of a retro-Cope procedure in natural product synthesis *i.e.* (+)-trianthine and (+)-iycorane (Scheme 5.39).

The synthesis of aspidofractine and kopsinine indole alkaloids reported by Wu et al. [86] included a rich source of appealing targets that form a test base for novel synthetic approaches. They selected aspidofractine and kopsinine as synthetic targets to exhibit their organocatalytic process (Scheme 5.40). The starting material was reacted with propynal in this exquisite Michael addition/aza-Michael addition/cyclization cascade reaction under catalysis to provide the key tetracyclic directly, presumably through the intermediacy of intermediates. In the second catalysis, the conjugate addition of formed compound to acrolein proceeded to furnish, after reduction, the dienamine which underwent [4+2]-cycloaddition/desulfurization sequence. After construction of the last ring by nucleophilic substitution,

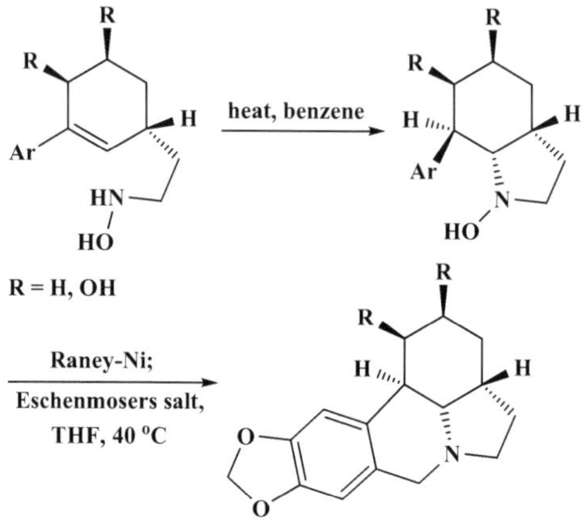

Scheme 5.39

an overall formation of kopsinine was carried out in a concise manner. The endgame *N*-formylation of kopsinine delivered aspidofractine [87].

Kametani et al. [88] have developed a new methodology including ring-expansion of an aziridinium ion for the construction of isopavine skeleton (Scheme 5.41). The synthesis began with the condensation of starting compound with NH$_2$OH, and subsequent reduction with Ra-Ni provided primary amine. A three-step formylation, reduction, and formylation sequence delivered formamide in 58% yield. A Bischler–Napieralski cyclization provided dihydroisoquinoline, which was reacted with CH$_3$I in Et$_2$O to produce the iminium ion. The aziridinium salt was obtained in high yield when iminium ion was reacted with ethereal CH$_2$N$_2$. The reframidine was formed in 20% yield over two steps when the crude aziridinium salt was reacted immediately with 6 N hydrochloric acid.

The synthesis started with the expansion of oxazolidinone. The requisite aldehyde, for the organocatalytic α-amination, was accessed from (+)-citronellal. The acetalization of aldehyde moiety in citronellal was followed by an oxidative cleavage of residual olefin function (Scheme 5.42). An enantioselective amination of aldehyde can be achieved by the reaction of aldehyde with dibenzyl azidocarboxylate using a proline-derived catalyst. The reduction of resultant intermediate was followed by treatment with potassium carbonate to synthesize the oxazolidinone. The reductive cleavage of nitrogen–nitrogen bond in hydrazide occurred by a sequential reduction procedure, using hydrogen/palladium followed by Ra-Ni, to

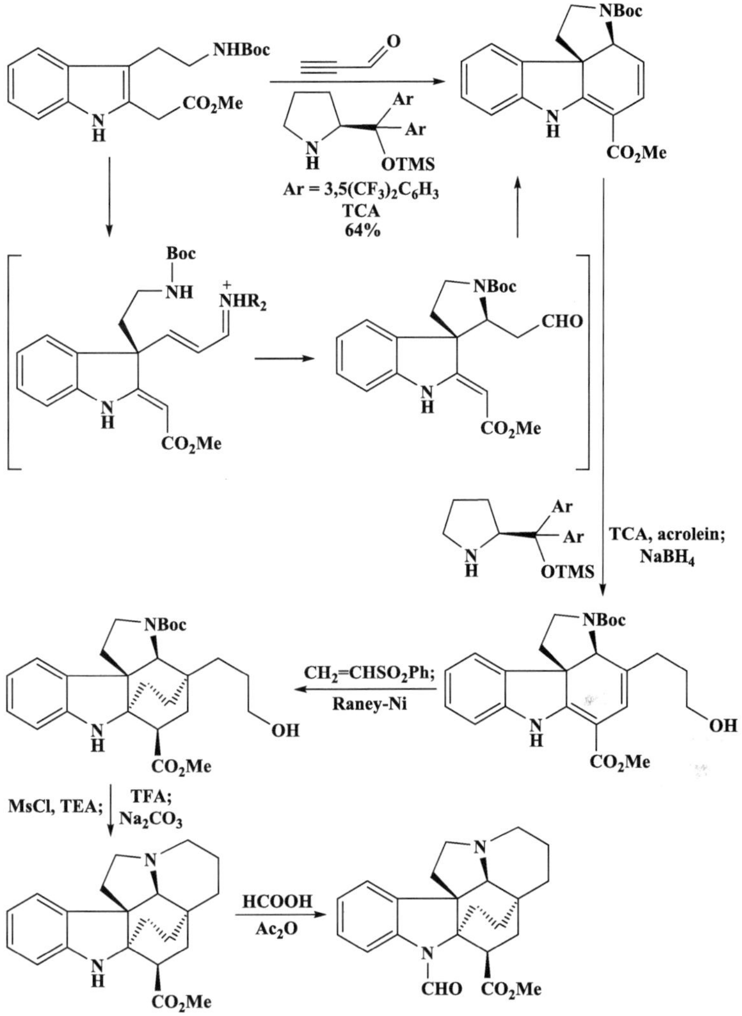

Scheme 5.40

deliver the oxazolidinone. The oxazolidinone was further reacted with p-toluenesulfonic acid to affect the desired cyclization. The reaction of formed aminoacetal with allyl trimethylsilane and titanium tetrachloride delivered allylated aminoacetal as a solo diastereomer [89].

The 9-hydroxyquinolizinone was hydrogenated utilizing Ra-Ni at 180°C and 250 atmospheres of H_2 to provide the cyclized bilactam, which was reduced with $LiAlH_4$ to provide the natural product (Scheme 5.43) [90].

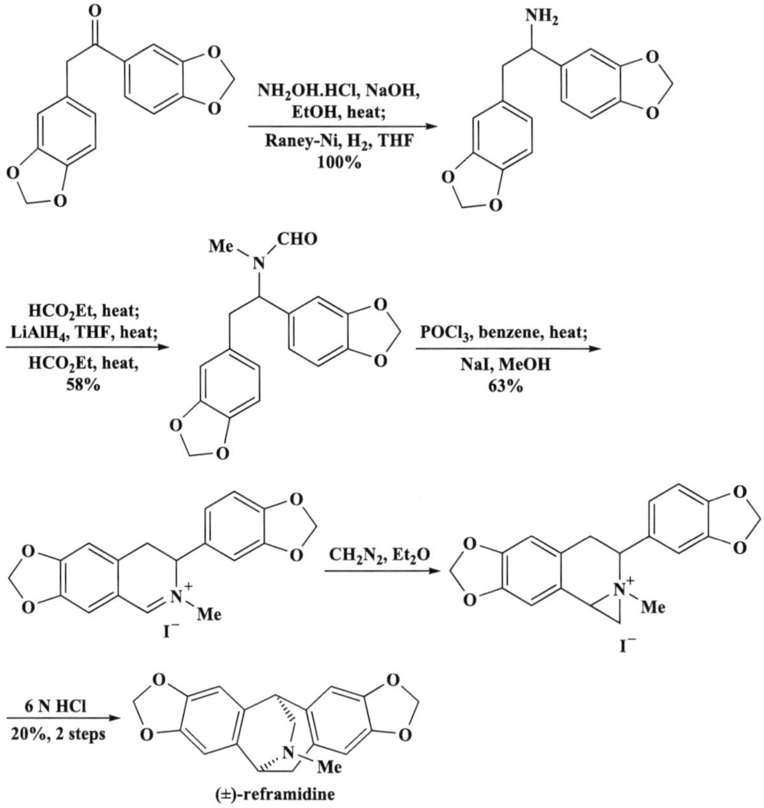

Scheme 5.41

The first de novo asymmetric overall synthesis of (+)-sparteine, the naturally occurring but rarely encountered enantiomer of the more common and commercially accessible (−)-sparteine, was based on the rearrangement of nitrone. The starting compound was optically pure (+)-norbornane-2,5-dione, which was converted into ketoazide in six steps. The tricyclic quinolizidine lactam was synthesized by an intramolecular Schmidt reaction in the presence of TiCl$_4$ with the immediate elimination of the ketal protecting group, which was further transformed into 4-chlorobutyl derivative utilizing standard processes. Although the azide was synthesized from this intermediate, a second intramolecular Schmidt reaction could never be achieved probably due to the preferential coordination of the tertiary amine to protic or Lewis acids. As a variant, the deprotection of hydroxylamine derivative afforded nitrone, which gave (+)-sparteine by

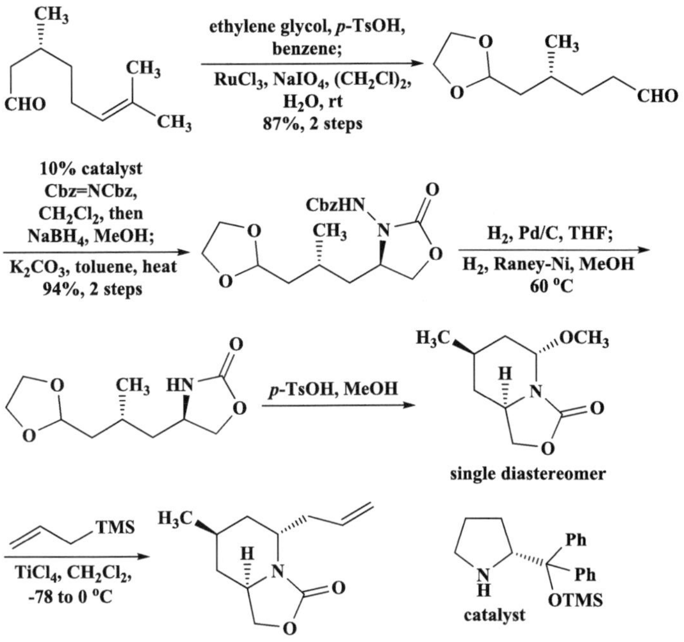

Scheme 5.42

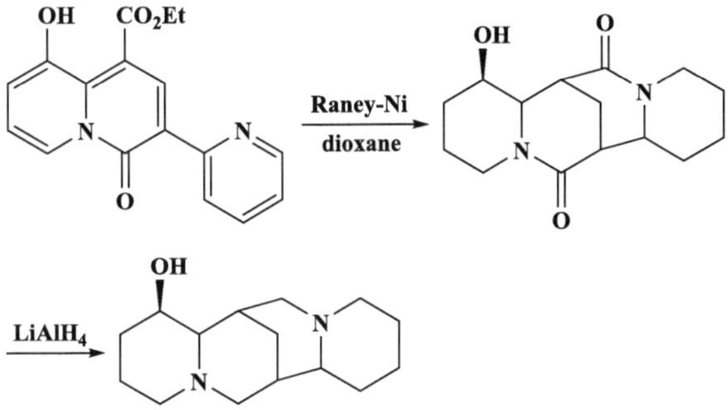

Scheme 5.43

photochemical Beckmann rearrangement to a tetracyclic lactam followed by standard reduction of the lactam function (Scheme 5.44) [91].

Total asymmetric syntheses of (−)-leuconicines A and B were accomplished utilizing a one-pot spirocyclization/intramolecular aza-Baylis–Hillman reaction a domino acylation/Knoevenagel cyclization to construct the F-ring, and an intramolecular Heck cyclization (Scheme 5.45) [92,93].

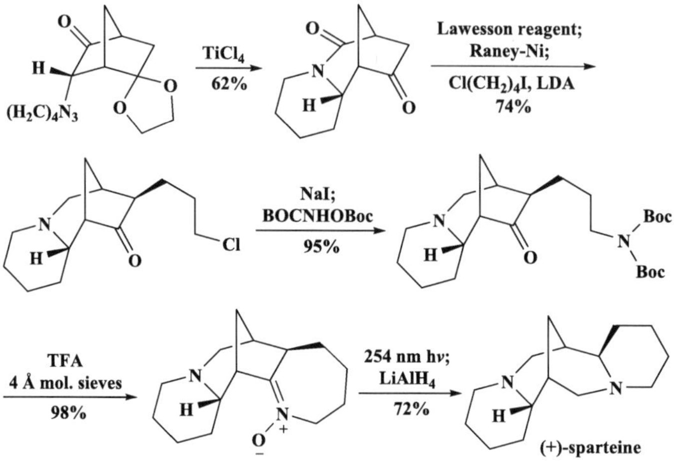

194　Raney nickel-assisted synthesis of heterocycles

Scheme 5.44

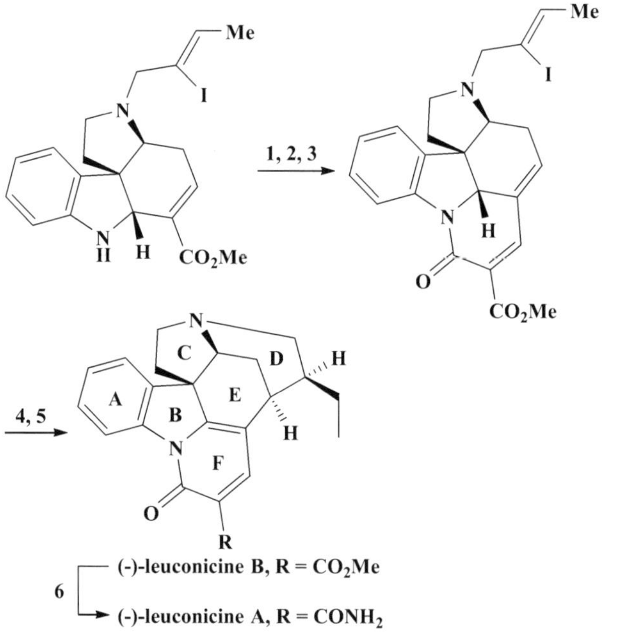

(+)-sparteine

(−)-leuconicine B, R = CO$_2$Me
(−)-leuconicine A, R = CONH$_2$

Reagents and conditions: (1) LiNMe(OMe), (2) DIBAH, (3) ClCOCH$_2$-CO$_2$Me, Et$_3$N, CH$_2$Cl$_2$, reflux, (4) Pd(OAc)$_2$, PPh$_3$, Et$_3$N, (5) Raney-Ni, (6) Me$_3$Al, NH$_3$.

Scheme 5.45

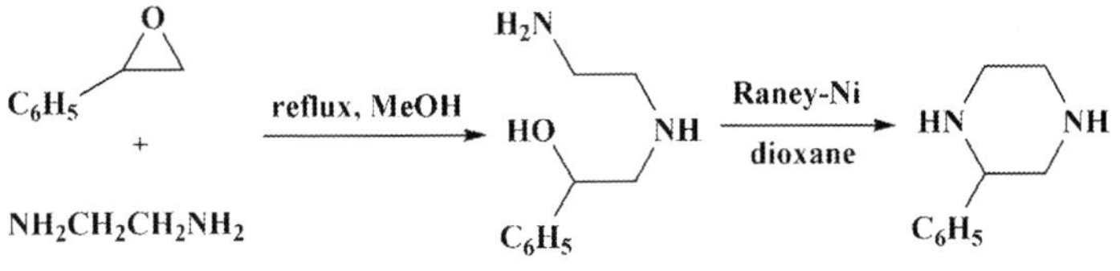

Scheme 5.46

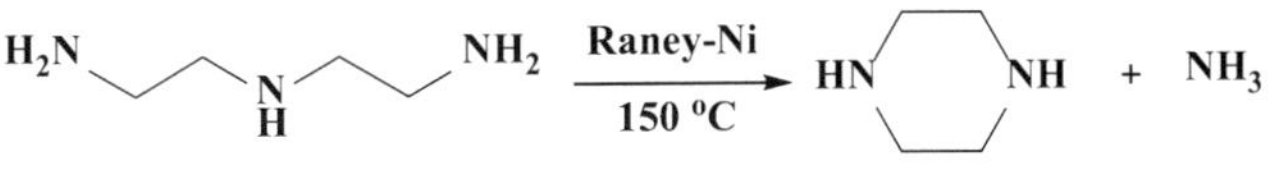

Scheme 5.47

5.10 Synthesis of six-membered *N,N*-heterocycles

Pollard and Kitchen [94] prepared 2-phenylpiperazine starting with the formation of 2-(2-aminoethylamino)-1-phenyl-ethanol by the reaction of 2-phenyl-oxirane with ethylenediamine in CH_3OH under reflux temperature, followed by reduction of 2-(2-aminoethylamino)-1-phenyl-ethanol with Ra-Ni (Scheme 5.46) [95].

The biological importance of piperazines has evoked considerable attention; intensive research has been conducted into general methodologies for the formation of piperazine core with increasing potential for uses in biological systems [96]. A variety of examples existed for the formation of substituted piperazine that included the construction of ring system and its derivatives by several approaches where most of them depend upon cyclization processes. For example, the piperazine ring was synthesized by heating diethylenetriamine with Ra-Ni under high temperature of about 150°C, with the liberation of NH_3 (Scheme 5.47) [95,97].

Different *N*-containing heterocycles were synthesized from amino alcohol derivatives as starting compounds and Ra-Ni as catalyst under H_2 atmosphere. For example, the 1-aminopropan-2-ol underwent a double hydrogen autotransfer to afford a mixture of *cis/trans*-piperazine and the related aromatic pyrazine [98]. The careful study of reaction conditions allowed the synthesis of aliphatic heterocyclic compounds as the key product (Scheme 5.48) [99].

The 5-methyl-4-nitro-2,1,3-benzoselenadiazole was transformed into 1,2,5-selenadiazolo[3,4-*g*]indole by Batcho–Leimgruber indole synthesis. The subsequent deselenation provided 6,7-diaminoindole, which on reaction with biacetyl provided 2,3-dimethylpyrrolo[2,3-*f*]quinoxaline

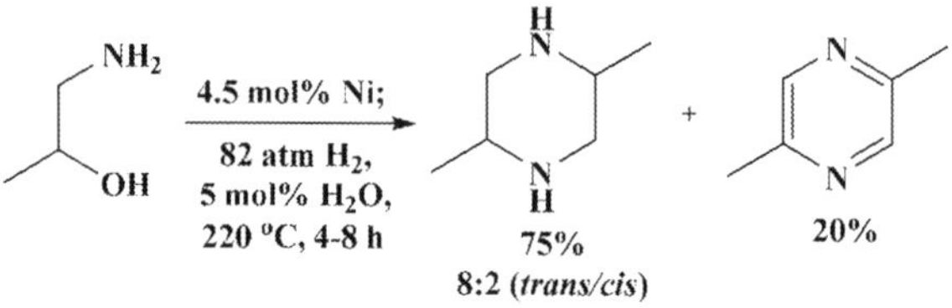

Scheme 5.48

in 80% yield from 1,2,5-selenadiazolo[3,4-g]indole. Batcho–Leimgruber indole synthesis on 5-methyl-4-nitro-2,1,3-benzoselenadiazole utilizing DMF-DMA in MeCN or dimethylformamide provided *trans*-isomer ($J = 13.1$ Hz) of enamine in 84% isolated yield. The reductive cyclization is generally performed with H_2 over a Pd catalyst or with Ra-Ni and NH_2NH_2 [100]. Efforts to convert enamine into 1,2,5-selenadiazolo[3,4-g]indole by hydrogenation over 10% palladium–carbon in tetrahydrofuran, at ambient conditions and up to 50 psi, provided only intact enamine. Further, no traces of 1,2,5-selenadiazolo[3,4-g]indole were detected in the dark reaction mixtures obtained when H_2 and Ra-Ni in tetrahydrofuran and/or alcohols were used at various temperatures. The enamine was heated with $NH_2NH_2 \cdot H_2O$ and Ra-Ni in tetrahydrofuran and/or alcohols to provide the 1,2,5-selenadiazolo[3,4-g]indole in 10% yield. However, more efficient ring-closure to 1,2,5-selenadiazolo[3,4-g]indole was achieved by heating enamine with Fe powder in CH_3COOH and C_2H_5OH, conditions known to transform the 4-nitro-benzoselenadiazole into amino compound without substantial deselenation [101]. The deselenation of benzoselenadiazole was usually accomplished with HI in HCl [102–106], $SnCl_2$ and HCl [107], Zn and HCl [108] or by ammonium sulfide [109]. Due to the potential oligomerization of the 2-unsubstituted indole in acid [110], deselenation was attempted with ammonium sulfide in C_2H_5OH. In contrast to earlier efficient deselenations with this reagent, only a minor amount of 1,2,5-selenadiazolo[3,4-g]indole was deselenated even in higher boiling alcohols. However, the diaminoindole was obtained in high yield by heating 1,2,5-selenadiazolo[3,4-g]indole with $NH_2NH_2 \cdot H_2O$ and Ra-Ni. The crude 6,7-diaminoindole was treated with SeO_2 to provide the selenadiazoloindole, quantitatively and spontaneously as indicated by thin layer chromatography. Moreover, crude 6,7-diaminoindole was treated with diacetyl compound to deliver the apparently unknown pyrroloquinoxaline in 80% isolated yield from 1,2,5-selenadiazolo[3,4-g]indole. Other pyrrolo[2,3-f]quinoxalines

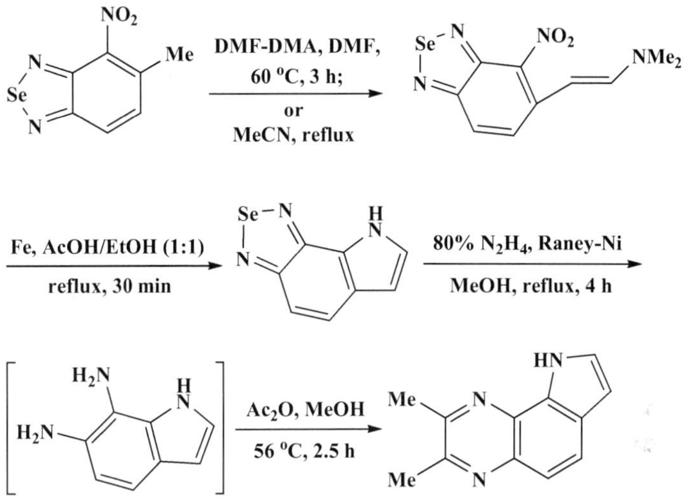

Scheme 5.49

have been synthesized and tested for biological activity. The sulfur analogue of 1,2,5-selenadiazolo[3,4-g]indole has been prepared by Fisher indole synthesis on 4-aminobenzothiadiazole (Scheme 5.49) [111–113].

5.11 Synthesis of six-membered *O*-heterocycles

Because of the simple fact that lactonization studies were carried out with MOM-protected α-hydroxyls, the hydroxyls of dithiane were protected as MOM-ethers [114]. The deprotection of dithiane occurred utilizing buffered I_2 in CH_3COCH_3-H_2O to yield the ketone [115]. The ketone was exposed to Ra-Ni under 40 atmospheric pressure of H_2 to deliver the carbinol. A variety of solvents and catalysts were screened to analyze the diastereoselectivity of the lactonization. The nonpolar solvent deuterated benzene delivered the best diastereoselectivity. The 1,5-diazabicyclo[4.3.0]non-5-ene (DBN) was proved to be the best catalyst (1:0; *cis*:*trans* at 70% conversion), only because 1,8-diazabicyclo[5.4.0]undec-7-ene (DBU) (10:1; *cis*:*trans*) often promoted the formation of major by-product (Scheme 5.50). The absolute rate of reaction turned out to be much slower in the hexyl series than it was with methyl ester series. The lactonization utilizing 1,5-diazabicyclo[4.3.0]non-5-ene seemed to slow down after about 75% conversion.

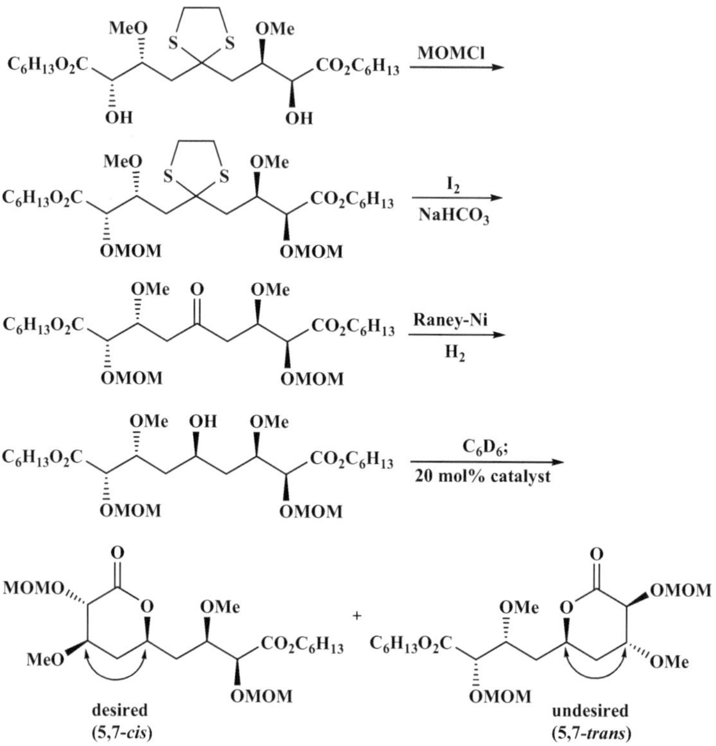

T. Ryba (2005): Kinetic lactonization ratios (hexyl)

catalyst	lactone ratio conv. (%)	time	*cis/trans*
DBU	100	3 d	10:1
TMG	<10	3 d	NO
DBN	70	3 d	1:0

Scheme 5.50

The formation of pyran fragment started with aldehyde, which on aldol condensation with chlorotitanium enolate of *N*-propionyl thiazolidinethione gave aldol product with excellent diastereoselectivity (98:2 diastereomeric ratio) [116]. The ester was formed by the elimination of chiral auxiliary followed by Wittig reaction. Hydrogenation of olefin and subsequent lactonization followed by reductive acetylation afforded acetate as a mixture of anomers (7:1). The acetate was transformed to pyran fragment in four steps (Scheme 5.51) [117].

The 3*H*-naphtho[2,1-*b*]pyrans with an amino or alkoxy residue in the 6-position displayed particularly high colorability. The 1-amino- and 1-alkoxy-3-hydroxynaphthalenes, required for the formation of 3*H*-naphtho[2,1-*b*]pyrans, were prepared from 2-naphthol (Scheme 5.52) [118].

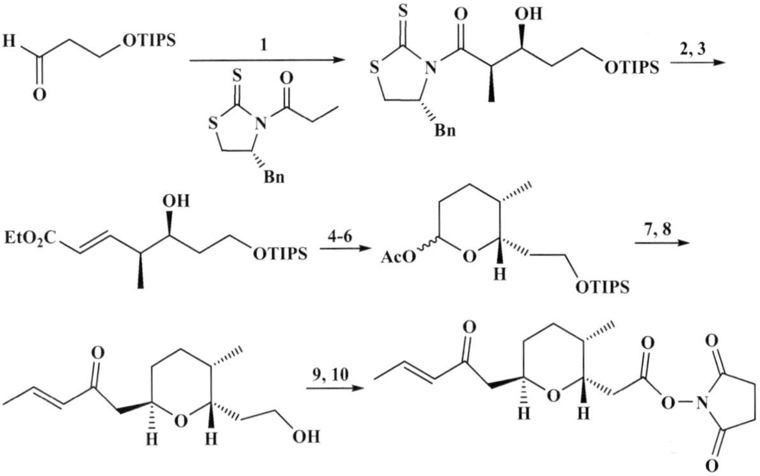

Reagents and conditions: (1) TiCl$_4$, NMP, (-)-sparteine, CH$_2$Cl$_2$, -78 °C, 87%, (2) *i*-Bu$_2$AlH, THF, -78 °C, (3) Ph$_3$P=CHCO$_2$Et, CH$_2$Cl$_2$, 78%, (4) H$_2$, Raney-Ni, EtOH, (5) PPTS, CH$_2$Cl$_2$, 40 °C, 81%, (6) *i*-Bu$_2$AlH, Py, DMAP, Ac$_2$O, CH$_2$Cl$_2$, -78 to -20 °C, 96%, (7) Et$_3$N, TMSOTf, 3-penten-2-one, CH$_2$Cl$_2$, 0 °C, then 78 °C, then acetate, 87%, 9:1 *dr*, (8) H$_2$SiF$_6$, MeCN, 0 °C, 75%, (9) H$_5$IO$_6$/CrO$_3$, MeCN, 77%, (10) *N*-hydroxysuccinimide, EDC.HCl, CH$_2$Cl$_2$, 100%

Scheme 5.51

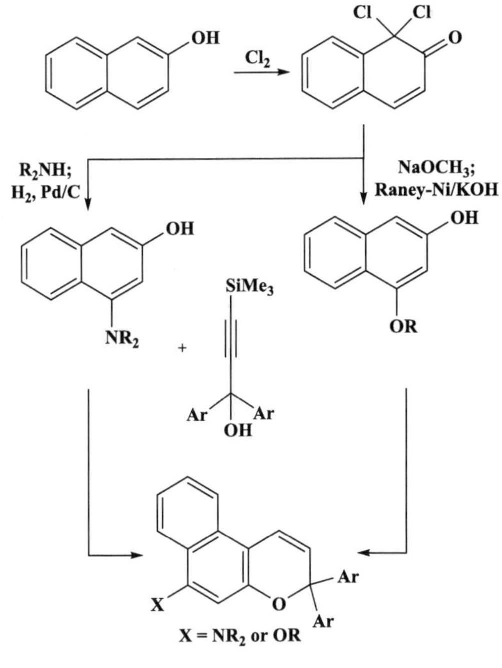

Scheme 5.52

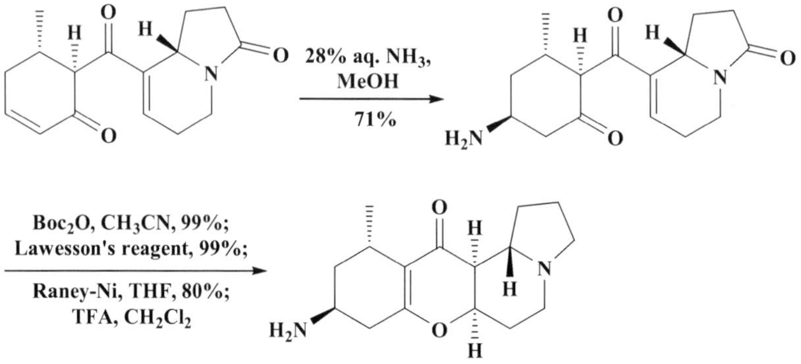

Scheme 5.53

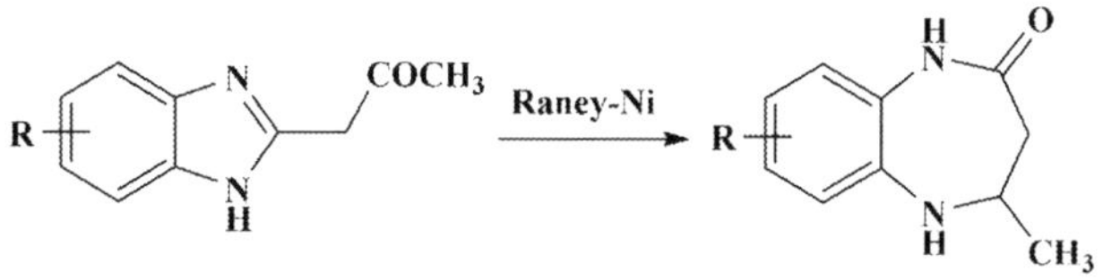

Scheme 5.54

The grandisine F was constructed by the reaction of diketone with NH_3 solution to give the tetracyclic amine as a solo isomer. Additional four steps to reduce the amide carbonyl group provided grandisine F (Scheme 5.53) [119].

5.12 Synthesis of seven-membered heterocycles

The *o*-nitro acetoacetanilide was reduced with Fe in alcoholic HCl to provide the 2-acetonyl benzimidazole. The catalytic hydrogenation of anilides in the presence of Ra–Ni afforded diazepinones (Scheme 5.54) [120].

A tricyclic system was prepared as shown in Scheme 5.55. The phthalimidomethylfurans were reacted with Br_2 in CH_3OH to provide the dihydrofurans. Hydrolysis followed by hydrogenation over Ra–Ni provided 1,4-diketones. The condensation of 2-amino-3-benzoylthiophenes and 1,4-diketones afforded pyrrolylthiophenes, which were fully characterized. The

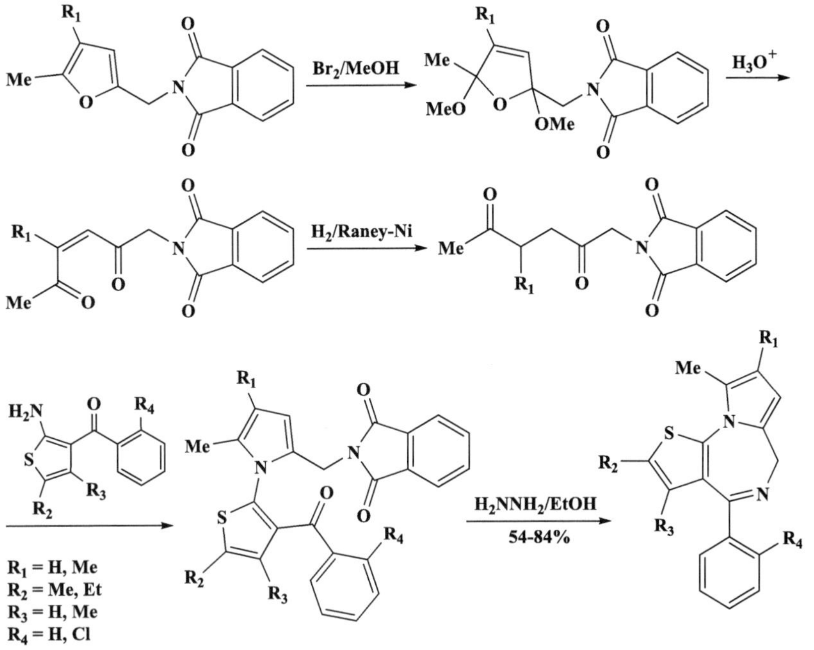

Scheme 5.55

removal of phthaloyl groups and ring-closure with NH$_2$NH$_2$ in refluxing C$_2$H$_5$OH was accompanied by ring-closure to diazepines [121].

Corey et al. [122] reported the first total synthesis of (+/−)-erythronolide B. The formation of *syn,syn,syn*-stereotetrad started from dienone (*rac*) by hydroboration followed by oxidation, to synthesize the dienone acid [123]. Then, treatment with Br$_2$/KBr solution provided bromo lactone, which was further transformed to epoxy acid under basic reaction conditions. The epoxy acid was then transformed to bromo epoxy lactone, from which the bromine was cleaved away through a radical reaction to synthesize the epoxy lactone. The epoxide was opened and the ketone was further stereoselectively reduced with Ra-Ni. Then, the OH groups were protected to give the dibenzoate, which already possessed the all *syn*-stereochemistry. Finally, the dibenzoate was converted to lactone, which was one of the key intermediate in the first total synthesis of erythronolide A (Scheme 5.56).

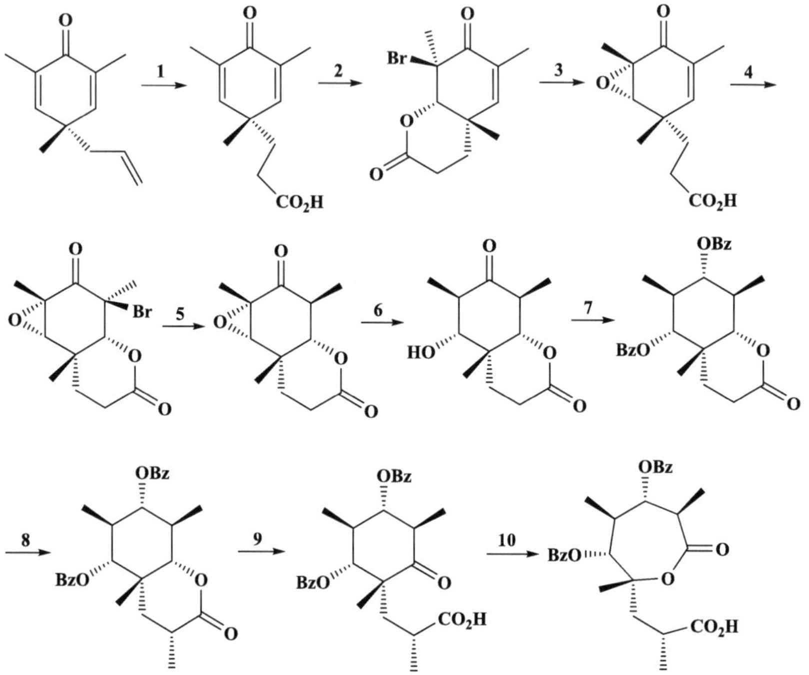

Reagents and conditions: (1) a) B_2H_6, THF, 0 to 10 °C, b) Jones chromic acid, 0 to -10 °C, (2) Br_2, KBr, H_2O, (3) aq. KOH, THF, (4) Br_2, KBr, H_2O, (5) N-Bu_3SnH, AIBN, PhH, (6) Al/Hg, THF, H_2O, 0 to -10 °C, (7) H_2, Raney-Ni, DME, -20 °C; BzCl, Py, (8) LDA, THF, -78 °C, then MeI, HMPA, -78 to > -45 °C, (9) a) LiOH, H_2O, b) CrO_3, H_2SO_4, acetone, -10 °C, (10) $MeCO_3H$, EtOAc, 55 °C.

Scheme 5.56

References

[1] (a) Balban AT, Oniciu DC, Katritzky AR. Aromaticity as a cornerstone of heterocyclic chemistry. Chem Rev 2004;104:2777–812; (b) Kaur N. Methods for metal and non-metal-catalyzed synthesis of six-membered oxygen-containing polyheterocycles. Curr Org Synth 2017;14:531–56; (c) Kaur N. Photochemical reactions: synthesis of six-membered N-heterocycles. Curr Org Synth 2017;14:972–98; (d) Kaur N, Dwivedi J, Kishore D. Solid-phase synthesis of nitrogen containing five-membered heterocycles. Synth Commun 2014;44:1671–729; (e) Dwivedi J, Kaur N, Kishore D, Kumari S, Sharma S. Synthetic and biological aspects of thiadiazoles and their condensed derivatives: an overview. Curr Top Med Chem 2016;16:2884–920.

[2] (a) Martins MAP, Cunico W, Pereira CMP, Flores AFC, Bonacorso HG, Zanatta N. 4-Alkoxy-1,1,1-trichloro-3-alken-2-ones: preparation and applications in heterocyclic synthesis. Curr Org Synth 2004;1:391–403; (b) Kaur N. Benign approaches for the microwave-assisted synthesis of five-membered 1,2-N,N-heterocycles. J Heterocycl Chem 2015;52:953–73; (c) Kaur N. Ionic liquids: promising but challenging solvents for the synthesis of N-heterocycles. Mini Rev Org Chem 2017;14:3–23; (d) Kaur N, Kishore D. An insight into hexamethylenetetramine: a versatile reagent in organic synthesis. J Iran Chem Soc 2013;10:1193–228; (e) Kaur N. An insight into medicinal

and biological significance of privileged scaffold: 1,4-benzodiazepine. Int J Pharm Biol Sci 2013;4:318–37.
[3] (a) Druzhinin SV, Balenkova ES, Nenajdenko VG. Recent advances in the chemistry of α,β-unsaturated trifluoromethylketones. Tetrahedron 2007;63:7753–808; (b) Kaur N. Metal catalysts for the formation of six-membered N-polyheterocycles. Synth React Inorg Met Org Nano Met Chem 2016;46:983–1020; (c) Kaur N. Applications of gold catalysts for the synthesis of five-membered O-heterocycles. Inorg Nano Met Chem 2017;47:163–87; (d) Kaur N. Solid-phase synthetic approach to the synthesis of azepine heterocycles of medicinal interest. Int J Pharm Biol Sci 2013;4:357–72; (e) Tyagi R, Kaur N, Singh B, Kishore D. A noteworthy mechanistic precedence in the exclusive formation of one regioisomer in the Beckmann rearrangement of ketoximes of 4-piperidones annulated to pyrazolo-indole nucleus by organocatalyst derived from TCT and DMF. Synth Commun 2013;43:16–25.
[4] (a) Lipshutz BH. Five-membered heteroaromatic rings as intermediates in organic synthesis. Chem Rev 1986;86:795–819; (b) Kaur N. Ruthenium catalysis in six-membered O-heterocycles synthesis. Synth Commun 2018;48:1551–87; (c) Kaur N. Palladium-catalyzed approach to the synthesis of five-membered O-heterocycles. Inorg Chem Commun 2014;49:86–119.
[5] (a) Wong HNC, Yu P, Yick CY. The use of furans in natural product syntheses. Pure Appl Chem 1999;71:1041–4; (b) Kaur N, Kishore D. Nitrogen-containing six-membered heterocycles: solid-phase synthesis. Synth Commun 2014;44:1173–211; (c) Kaur N, Kishore D. Solid-phase synthetic approach toward the synthesis of oxygen-containing heterocycles. Synth Commun 2014;44:1019–42.
[6] (a) Rassu G, Zanardi F, Battistini L, Casiraghi G, Rassu G, Zanardi F, Battistini L, Casiraghi G. The synthetic utility of furan-, pyrrole- and thiophene-based 2-silyloxy dienes. Chem Soc Rev 2000;29:109–18; (b) Kaur N. Microwave-assisted synthesis of five-membered O-heterocycles. Synth Commun 2014;44:3483–508; (c) Kaur N. Microwave-assisted synthesis of five-membered O,N-heterocycles. Synth Commun 2014;44:3509–37.
[7] (a) Chinchilla R, Najera C, Yus M. Metalated heterocycles and their applications in synthetic organic chemistry. Chem Rev 2004;104:2667–722; (b) Kaur N. Microwave-assisted synthesis of five-membered O,N,N-heterocycles. Synth Commun 2014;44:3229–47; (c) Kaur N. Synthesis of six- and seven-membered heterocycles under ultrasound irradiation. Synth Commun 2018;48:1235–58.
[8] (a) Lee HK, Chan KF, Hui CW, Yim HK, Wu XW, Wong HNC. Use of furans in synthesis of bioactive compounds. Pure Appl Chem 2005;77:139–43; (b) Kaur N. Metal catalysts: applications in higher-membered N-heterocycles synthesis. J Iran Chem Soc 2015;12:9–45; (c) Kaur N. Insight into microwave-assisted synthesis of benzo derivatives of five-membered N,N-heterocycles. Synth Commun 2015;45:1269–300.
[9] (a) Schröter S, Stock C, Bach T. Regioselective cross-coupling reactions of multiple halogenated nitrogen-, oxygen-, and sulfur-containing heterocycles. Tetrahedron 2005;61:2245–67; (b) Isambert N, Lavilla R. Heterocycles as key substrates in multicomponent reactions: the fast lane towards molecular complexity. Chem Eur J 2008;14:8444–54.
[10] (a) Dondoni A. Heterocycles in organic synthesis: thiazoles and triazoles as exemplar cases of synthetic auxiliaries. Org Biomol Chem 2010;8:3366–85; (b) Kaur N. Synthesis of fused five-membered N,N-heterocycles using microwave irradiation. Synth Commun 2015;45:1379–410; (c) Kaur N. Microwave-assisted synthesis of seven-membered S-heterocycles. Synth Commun 2014;44:3201–28.
[11] Billica HR and Adkins H. Organic synthesis. E.C. Horning (Ed.). John Wiley & Sons: New York, 1955;3:176–180.

[12] Mozingo R. Organic synthesis. E.C. Horning (Ed.). John Wiley & Sons: New York, 1955;3:181–183.
[13] Plieninger H, Decker M. Eine neue synthese für pyrrolone, insbesondere für "isooxyopsopyrrol" und "isooxyopsopyrrol-carbonsäure". Liebigs Ann Chem 1956;598:198–207.
[14] Plieninger H, Kune J. Synthese der "oxyopsopyrrolcarbonsäure" und weitere untersuchungen in der pyrrolon-reihe. Liebigs Ann Chem 1964;680:60–9.
[15] Gossauer A, Weller J-P. Synthesen von gallenfarbstoffen, VI: total synthese des (+)(4R,16R)- und (-)(4R,16S)-[18-vinyl]mesourobilin IXα-dimethylesters. Chem Ber 1978;111:486–501.
[16] Schoenleber RW, Kim Y, Rapoport H. Relative stereochemistry of the A ring of plant bile pigments. J Am Chem Soc 1984;106:2645–51.
[17] Boiadjiev SE, Lightner DA. Dipyrrinones-constituents of the pigments of life. A review. Org Prep Proced Int 2006;38:347–99.
[18] Montforts F-P, Schwartz UM. Totalsynthese von (±)-bonellin-dimethylester. Liebigs Ann Chem 1991;8:709–25.
[19] Yakovlev KV, Petrov DV, Dokichev VA, Tomilov YV. Catalytic synthesis of 1,3-propylenediamines. Russ J Org Chem 2011;47:168–72.
[20] Bocchi V, Palla G. Synthesis and spectroscopic characteristics of 2,3-biindolyl and 2,2-indolylpyrroles. Tetrahedron 1984;40:3251–6.
[21] Guo X, Peng Z, Jiang S, Shen J. Convenient and scalable process for the preparation of indole via Raney nickel-catalyzed hydrogenation and ring closure. Synth Commun 2011;41:2044–52.
[22] Farmer SC, Berg SH. Ring contracting sulfur extrusion from oxidized phenothiazine ring systems. Molecules 2008;13:1345–52.
[23] Ma D, Pu X, Wang J. Efficient formal synthesis of the dendrobatid alkaloid, indolizidine (-)-209B. Tetrahedron: Asymmetry 2002;13:2257–60.
[24] Michael JP, Gravestock D. Vinylogous urethanes in alkaloid synthesis. Applications to the synthesis of racemic indolizidine 209B and its (5R*,8S*,8aS*)-(±) diastereomer, and to (-)-indolizidine 209B. J Chem Soc, Perkin Trans 1 2000;12:1919–28.
[25] Herberhold M, Seela F, Richter R. Aminomethylierung von 3,7-dihydropyrrolo[2,3-d]pyrimidinen an C-5 - ein weg zur synthese von aglycon-analoga des nucleosids "Q". Chem Ber 1978;111:2925–30.
[26] Davoll J. Pyrrolo[2,3-d]pyrimidines. J Chem Soc 1960;0:131–8.
[27] Kita Y, Gotanda K, Fujimori C, Murata K, Wakayama R, Matsugi M. Polonovski-type reaction induced by O-silylated ketene acetals. J Org Chem 1997;62:8268–70.
[28] Shih H, Cottam HB, Carson DA. Facile synthesis of 9-substituted 9-deazapurines as potential purine nucleoside phosphorylase inhibitors. Chem Pharm Bull 2002;50:364–367.
[29] El-Salfiti MK. PhD Thesis. University of Toronto; 2012.
[30] Denmark SE, Schnute ME, Marcin LR, Thorarensen A. Nitroalkene inter [4+2]/intra [3+2] tandem cycloadditions. 7. Application of (R)-(-)-2,2-diphenylcyclopentanol as the chiral auxiliary. J Org Chem 1995;60:3205–20.
[31] Bonjoch J, Catena J, Valls N. Total synthesis of (±)-deethylibophyllidine: studies of a Fischer indolization route and a successful approach via a Pummerer rearrangement/thionium ion-mediated indole cyclization. J Org Chem 1996;61:7106–15.
[32] Kozikowski AP, Greco MN, Springer J. Total synthesis of the unique mycotoxin alphacyclopiazonic acid (alpha-CA): an unusual dimethyl zinc mediated replacement of a phenylthio substituent by a methyl group and a contrathermodynamic Raney nickel desulfurization reaction. J Am Chem Soc 1984;106:6873–4.
[33] Israel M, Jones LC. On the reactions of β-ketoesters with 2,3-diaminopyridine and its derivatives. J Heterocycl Chem 1973;10:201–7.

[34] Savelli F, Boido A, Mule A, Piu L, Alamanni MC, Pirisino G, Satta M, Peana A. 1,4-Disubstituted 1,3-dihydro-2*H*-1,5-benzo- and chlorobenzodiazepin-2-ones with activity on the central nervous system. Farmaco 1989;44:125–40.
[35] Savelli F, Boido A, Satta M, Peana A, Marzano C. Synthesis and CNS activities of pyridopyrazinone and pyridodiazepinone derivatives. IL FARMACO 1994;49:259–65.
[36] Kim T, Kim K, Park YJ. A novel method for the synthesis of 2,3-benzo-1,3a,6a-triazapentalenes through Pummerer-type reactions of γ-(benzotriazol-1-yl)allylic sulfoxides. Eur J Org Chem 2002;3:493–502.
[37] List B. Direct catalytic asymmetric α-amination of aldehydes. J Am Chem Soc 2002;124:5656–7.
[38] Kumaragurubaran N, Juhl K, Zhuang W, Bøgevig A, Jørgensen KA. Directly proline-catalyzed asymmetric α-amination of ketones. J Am Chem Soc 2002;124:6254–5.
[39] Trost BM, Dake GR. Nucleophilic α-addition to alkynoates. A synthesis of dehydroamino acids. J Am Chem Soc 1997;119:7595–6.
[40] Zhong M, Nowak I, Cannon JF, Robins MJ. Structure and synthesis of 6-(substituted-imidazol-1-yl)purines: versatile substrates for regiospecific alkylation and glycosylation at N9I. J Org Chem 2006;71:4216–21.
[41] Zhong M, Nowak I, Robins MJ. Regiospecific and highly stereoselective coupling of 6-(substituted-imidazol-1-yl)purines with 2-deoxy-3,5-di-*o*-(*p*-toluoyl)-α-D-*erythro*-pentofuranosyl chloride. Sodium-salt glycosylation in binary solvent mixtures: improved synthesis of cladribine. J Org Chem 2006;71:7773–9.
[42] Rodriguez JR, Rumbo A, Castedo L, Mascarenas JL. [5+2] Pyrone-alkene cycloaddition approach to tetrahydrofurans. Expeditious synthesis of (±)-nemorensic acid. J Org Chem 1999;64:4560–3.
[43] Bowden K, Heilbron IM, Jones ERH. Researches on acetylenic compounds. Part I. The preparation of acetylenic ketones by oxidation of acetylenic carbinols and glycols. J Chem Soc 1946;0:39–45.
[44] Noe CR, Knollmuller M, Dungler K, Gartner P. Pheromones, 2. Mitt. A process for the preparation of ketones and spiroketals by desulfurization of thienyl ethers. Monatsh Chem 1991;122:185–94.
[45] Rentner J, Kljajic M, Offner L, Breinbauer R. Recent advances and applications of reductive desulfurization in organic synthesis. Tetrahedron 2014;70:8983–9027.
[46] Wang Y, Dai W-M. Total synthesis of diastereomeric marine butenolides possessing a *syn*-aldol subunit at C10 and C11 and the related C11-ketone. Tetrahedron 2010;66:187–96.
[47] Briot A, Bujard M, Gouverneur V, Nolan SP, Mioskowski C. Improvement in olefin metathesis using a new generation of ruthenium catalyst bearing an imidazolylidene ligand: synthesis of heterocycles. Org Lett 2000;2:1517–19.
[48] Horsfall LR. PhD Thesis. London: University College; 2011.
[49] Bogevig A, Juhl K, Kumaragurubaran N, Zhuang W, Jørgensen KA. Direct organocatalytic asymmetric α-amination of aldehydes - a simple approach to optically active α-amino aldehydes, α-amino alcohols, and α-amino acids. Angew Chem, Int Ed 2002;41:1790–3.
[50] Tomita F, Takahashi K, Tomaoki T. Quinocarcin, a novel antitumor antibiotic. 3. Mode of action. J Antibiot 1984;37:1268–72.
[51] Garner P, Sunitha K, Shanthilal T. An approach to the 3,8-diazabicyclo[3.2.1]octane moiety of naphthyridinomycin and quinocarcin via 1,3-dipolar cycloaddition of photochemically generated azomethine ylides. Tetrahedron Lett 1988;29:3525–8.
[52] Tereshima S, Saito S, Tamura O, Kobayashi Y, Matsuda F, Katoh T. Synthetic studies on quinocarcin and its related compounds. 1. Synthesis of enantiomeric pairs of the ABE ring systems of quinocarcin. Tetrahedron 1994;50:6193–208.

[53] Williams RM, Scott JD. Chemistry and biology of the tetrahydroisoquinoline antitumor antibiotics. Chem Rev 2002;102:1669–730.
[54] Owen LN, Peto AG. Alicyclic glycols. Part XIII. 1: 2-Bishydroxymethylcyclopentane. J Chem Soc 1955;0:2383–90.
[55] Craig D, Meadows JD, Pécheux M. Heterocyclic synthesis by C-C bond formation. Thionium ion-mediated preparation of substituted pyrrolidines and piperidines. Tetrahedron Lett 1988;39:147–50.
[56] Mitchinson A, Nadin A. Saturated nitrogen heterocycles. J Chem Soc, Perkin Trans 1 1999;18:2553–81.
[57] Ibuka T, Minakata H, Mitsui Y, Hayashi K, Taga T, Inubushi Y. New synthetic routes to (+)-perhydrohistrionicotoxin. Stereoselective synthesis of (6S^*,7S^*,8S^*)-7-butyl-8-hydroxy-1-azaspiro(5.5)-undecan-2-one and its (6R^*)-isomer. Chem Pharm Bull 1982;30:2840–59.
[58] Wardrop DJ, Zhang W, Landrie CL. Stereoselective nitrenium ion cyclizations: asymmetric synthesis of the (+)-Kishi lactam and an intermediate for the preparation of fasicularin. Tetrahedron Lett 2004;45:4229–31.
[59] Sinclair A, Stockman RA. Thirty-five years of synthetic studies directed towards the histrionicotoxin family of alkaloids. Nat Prod Rep 2007;24:298–326.
[60] Duhamel P, Kotera M, Monteil T. New piperidinic synthons via ring contraction. Formal synthesis of (±)-perhydrohistrionicotoxin. Bull Chem Soc Jpn 1986;59: 2353–2355.
[61] Bringmann G, Jansen JR, Rink H-P. Regioselective and atropoisomeric-selective aryl coupling to give naphthyl isoquinoline alkaloids: the first total synthesis of (-)-ancistrocladine. Angew Chem, Int Ed 1986;25:913–15.
[62] Bobbitt J, Sih J. Synthesis of isoquinolines. VII. 4-Hydroxy-1,2,3,4-tetrahydroisoquinolines. J Org Chem 1968;33:856–8.
[63] Gensler WJ. Organic reactions, 6. New York: Wiley; 1951. R. Adams p. 191–206.
[64] Casiraghi G, Zhanardi F, Rassu G, Spanu P. Stereoselective approaches to bioactive carbohydrates and alkaloids - with a focus on recent syntheses drawing from the chiral pool. Chem Rev 1995;95:1677–716.
[65] Gebarowski P, Sas W. Asymmetric synthesis of novel polyhydroxylated derivatives of indolizidine and quinolizidine by intramolecular 1,3-dipolar cycloaddition of N-(3-alkenyl)nitrones. Chem Commun 2001;10:915–16.
[66] Boekelheide V, Rothchild S. Curariform activity and chemical structure. V.1 Syntheses in the quinolizidine series. J Am Chem Soc 1949;71:879–86.
[67] Bohlmann F, Ottawa N, Keller R. Aufbau des tetrahydro-chinolizons und des "bispidins" beiträge zur synthese des cytisins. Justus Liebigs Ann Chem 1954;587:162–76.
[68] Veerasamy N, Carlson EC, Carter RG. Expedient enantioselective synthesis of cermizine D. Org Lett 2012;14:1596–9.
[69] Veerasamy N, Carlson EC, Collett ND, Saha M, Carter RG. Enantioselective approach to quinolizidines: total synthesis of cermizine D and formal syntheses of senepodine G and cermizine C. J Org Chem 2013;78:4779–800.
[70] Carlson EC, Rathbone LK, Yang H, Collett ND, Carter RG. Improved protocol for asymmetric, intramolecular heteroatom Michael addition using organocatalysis: enantioselective syntheses of homoproline, pelletierine, and homopipecolic acid. J Org Chem 2008;73:5155–8.
[71] Bal BS, Childers WE, Pinnick HW. Oxidation of α,β-unsaturated aldehydes. Tetrahedron 1981;37:2091–6.
[72] Evans DA, Bartroli J, Shih TL. Enantioselective aldol condensations. 2. *Erythro*-selective chiral aldol condensations via boron enolates. J Am Chem Soc 1981;103:2127–9.

[73] Strelchyonok OA, Avvakumov GV, Akhrem AA. Pregnancy-associated molecular variants of human serum transcortin and thyroxine-binding globulin. Carbohydr Res 1984;134:133–40.
[74] Brown RT, Ford MJ, Wingfield M. Stereoselective total synthesis of (±)-hirsutine and related *Corynanthé* alkaloids. J Chem Soc, Chem Commun 1984;13:847–8.
[75] Wenkert E, Yashwant DV, Yadav JS. Short syntheses of hirsutine and geissoschizine. J Am Chem Soc 1980;102:7972–7972.
[76] Kim G, Chu-Moyer MY, Danishefsky SJ, Schulte GK. The total synthesis of indolizomycin. J Am Chem Soc 1993;115:30–9.
[77] Guarna A, Lombardi E, Machetti F, Occhiato EG, Scarpi D. Modification of the aza-Robinson annulation for the synthesis of 4-methyl-benzo[c]quinolizin-3-ones, potent inhibitors of steroid 5α-reductase. J Org Chem 2000;65:8093–5.
[78] Heathcock CH, Davidsen SK, Mills SG, Sanner MA. *Daphniphyllum* alkaloids. 10. Classical total synthesis of methyl homodaphniphyllate. J Org Chem 1992;57:2531–44.
[79] Mook JR, Lackey K, Bennett C. Synthesis of phenanthridin-3-one derivatives: nonsteroidal inhibitors of steroid 5-α-reductase. Tetrahedron Lett 1995;36:3969–72.
[80] Kim W-S, Kim H-J, Cho C-G. Regioselectivity in the Stille coupling reactions of 3,5-dibromo-2-pyrone. J Am Chem Soc 2003;125:14288–9.
[81] Blum J, Gelman D, Baidossi W, Shakh E, Rosenfeld A, Aizenshtat Z. Palladium-catalyzed methylation of aryl and vinyl halides by stabilized methyl aluminum and methyl gallium complexes. J Org Chem 1997;62:8681–6.
[82] Kim W-S, Lee J-H, Kang J, Cho C-G. Diels-Alder cycloadditions of 3-phenylamino-5-bromo-2-pyrone for the synthesis of constrained α-amino acid derivatives. Tetrahedron Lett 2004;45:1683–7.
[83] Meyers AI, Berney D. Asymmetric synthesis of the (4aS,8aR,8S)-hydrolilolidone system. A formal total synthesis of unnatural (+)-aspidospermine. J Org Chem 1989;54:4673–6.
[84] Cho H-K, Tam NT, Cho C-G. Total synthesis of (±)-aspidospermidine starting from 3-ethyl-5-bromo-2-pyrone. Bull Korean Chem Soc 2010;31:3382–4.
[85] Oppolzer W, Spivey AC, Bochet CG. Suprafaciality of thermal N-4-alkenylhydroxylamine cyclizations: syntheses of (+)-alpha-lycorane and (+)-trianthine. J Am Chem Soc 1994;116:3139–40.
[86] Wu X, Huang J, Guo B, Zhao L, Liu Y, Chen J, Cao W. Enantioselective Michael/aza-Michael/cyclization organocascade to tetracyclic spiroindolines: concise total synthesis of kopsinine and aspidofractine. Adv Synth Catal 2014;356:3377–82.
[87] Sun B-F. Total synthesis of natural and pharmaceutical products powered by organocatalytic reactions. Tetrahedron Lett 2015;56:2133–40.
[88] Kametani T, Hirata S, Ogasawara K. Studies on the syntheses of heterocyclic compounds. Part DXXVI. A novel synthesis of isopavine-type alkaloids. Total synthesis of (±)-reframidine. J Chem Soc Perkin Trans 1973;10:1466–70.
[89] Chandra A. PhD Thesis. Graduate School of Vanderbilt University; 2011.
[90] O'Neill W. PhD Thesis. The University of Edinburgh; 2009.
[91] Smith BT, Wendt JA, Aube J. First asymmetric total synthesis of (+)-sparteine. Org Lett 2002;4:2577–9.
[92] Sirasani G, Andrade RB. Total synthesis of (-)-leuconicine A and B. Org Lett 2011;13:4736–7.
[93] Ishikura M, Abe T, Choshi T, Hibino S. Simple indole alkaloids and those with a non-rearranged monoterpenoid unit. Nat Prod Rep 2013;30:694–752.
[94] Pollard CB and Kitchen LJ. 1946. US Patent, 2: 022.
[95] Al-Ghorbani M, Begum AB, Mamatha SV, Khanum SA. Piperazine and morpholine: synthetic preview and pharmaceutical applications. J Chem Pharm Res 2015;7:281–301.

[96] Kumar AC, Swamy SN, Thimmegowda NR, Prasad SB, Yip GW, Rangappa KS. Synthesis and evaluation of 1-benzhydryl-sulfonyl-piperazine derivatives as inhibitors of MDA-MB-231 human breast cancer cell proliferation. Med Chem Res 2007;16:179–87.
[97] Kitchen LJ, Pollard CB. Derivatives of piperazine. XXI. Synthesis of piperazine and C-substituted piperazines. J Am Chem Soc 1947;69:854–5.
[98] Langdon WK, Levis WW, Jackson DR. 2,5-Dimethylpiperazine synthesis from isopropanolamine. Ind Eng Chem Process Des Dev 1962;1:153–6.
[99] Guillena G, Ramon DJ, Yus M. Hydrogen autotransfer in the N-alkylation of amines and related compounds using alcohols and amines as electrophiles. Chem Rev 2010;110:1611–41.
[100] Clark RD, Repke DB. The Leimgruber-Batcho indole synthesis. Heterocycles 1984;22:195–221.
[101] Efros LS. The aromatic bond and some problems of the structure of aromatic compounds. Russ Chem Rev 1960;29:66–78.
[102] Tian W, Grivas S. A useful methodology for the synthesis of 2-methyl-4-nitrobenzimidazoles. Synthesis 1992;12:1283–6.
[103] Ronne E, Grivas S, Olsson K. Synthetic routes to the carcinogen IQ and related 3H-imidazo[4,5-f]quinolones. Acta Chem Scand 1994;48:823–30.
[104] Grivas S, Tian W. Convenient synthesis of 4-methoxy-3-nitro-1,2-benzenediamine and its reaction into 6-methoxy-5-nitroquinoxalines. Acta Chem Scand 1992;46:1109–13.
[105] Tian W, Grivas S. Synthesis of 6-halo-5-nitroquinoxalines. J Heterocycl Chem 1992;29:1305–8.
[106] Tian W, Grivas S, Olsson K. Nitration of 5-fluoro-2,1,3-benzoselenadiazoles, and the synthesis of 4-fluoro-3-nitro-, 4-fluoro-6-nitro-, 5-fluoro-3-nitro-o-phenylenediamines and 3,4-diamino-2-nitrophenols by subsequent deselenation. J Chem Soc, Perkin Trans 1 1993;2:257–61.
[107] Tsubata Y, Suzuki T, Miyashi T. Single-component organic conductors based on neutral radicals containing the pyrazino-TCNQ skeleton. J Org Chem 1992;57:6749–55.
[108] Sawicki E, Carr A. Structure of 2,1,3-benzoselenadiazole and its derivatives. I. Ultraviolet-visible absorption spectra. J Org Chem 1957;22:503–6.
[109] Grivas S, Tian W, Ronne E, Lindström S, Olsson K. Synthesis of mutagenic methyl- and phenyl-substituted 2-amino-3H-imidazo[4,5-f]quinoxalines via 2,1,3-benzoselenadiazoles. Acta Chem Scand 1993;47:521–8.
[110] Katritzky AR, Rees CW. Comprehensive heterocyclic chemistry, 4. Oxford: Pergamon Press; 1984. p. 206–8.
[111] Titov GA, Chetverikov VP, Bundel' YG, Malyuga OA, Ivchenko TI, Alekseeva EN. Synthesis and antimicrobial activity of [1,2,5]thiadiazolo[3,4-g]indoles. Pharm Chem J (Engl Transl) 1987;21:867–70.
[112] Chetverikov VP, Titov GA, Bundel' YG, Luk'yanova MS, Kurilenko VM. Synthesis and pharmacological activity of 3H-[1,4]diazepino[2,3-g]indoles. Pharm Chem J (Engl Transl) 1987;21:874–8.
[113] Edin M, Grivas S. First synthesis of 6,7-diaminoindole and 1,2,5-selenadiazolo[3,4-g]indole. ARKIVOC 2000(i):1–5.
[114] Amato JS, Karady S, Sletzinger M, Weinstock LM. A new preparation of chloromethyl methyl ether free of bis[chloromethyl] ether. Synthesis 1979;12:970–1.
[115] Nicolaou KC, Bunnage ME, McGarry DG, Shi SH, Somers PK, Wallace PA, Chu XJ, Agrios KA, Gunzner JL, Yang Z. Total synthesis of brevetoxin A: Part 1: first generation strategy and construction of BCD ring system. Chem Eur J 1999;5:599–617.

[116] Crimmins MT, King BW, Tabet EA, Chaudhary K. Asymmetric aldol additions: use of titanium tetrachloride and (-)-sparteine for the soft enolization of N-acyl oxazolidinones, oxazolidinethiones, and thiazolidinethiones. J Org Chem 2001;66:894–902.
[117] Raju BR, Saikia AK. Asymmetric synthesis of naturally occurring spiroketals. Molecules 2008;13:1942–2038.
[118] van Gemert B. Organic photochromic and thermochromic compounds. In: Crano JC, Guglielmetti RJ, editors. Main photochromic families. New York: Plenum Press; 1999. 1 p. 111–40.
[119] Kurasaki H, Okamoto I, Morita N, Tamura O. Total synthesis of grandisine D. Org Lett 2009;11:1179–81.
[120] Hornyna J. 1965. Czecgiskivakian Patent, 113: 422; Chem. Abstr. 63: 1812.
[121] Watthey JWH, Stanton J, Peet NP. Azepines, Part 2. John Wiley & Sons; 1984. A. Rosowsky. ISBN 0-471-01878-3 (v. 1), ISBN 0-471-89592-X (v. 2).
[122] Corey EJ, Kim S, Yoo S, Nicolaou KC, Melvin LS, Brunelle DJ, Falck JR, Trybulski EJ, Lett R, Sheldrake PW. Total synthesis of erythromycins. 4. Total synthesis of erythronolide B. J Am Chem Soc 1978;100:4620–2.
[123] Corey EJ, Trybulski EJ, Melvin LS, Secrist JA, Lett R, Sheldrake PW, Falck JR, Brunelle DJ, Haslanger MF, Kim S, Yoo S. Total synthesis of erythromycins. 3. Stereoselective routes to intermediates corresponding to C(1) to C(9) and C(10) to C(13) fragments of erythronolide B. J Am Chem Soc 1978;100:4618–20.

Conclusion

Heterocycles mimic the major pharmaceutical products and natural products with biological activities. An important part of biologically active compounds is formed by heterocycles. Other important practical uses of heterocycles are that they act as modifiers and additives in a wide range of industries such as reprography, plastics, cosmetics, vulcanization accelerators, solvents, antioxidants, and information storage.

It is easy to realize that why both the development of new approaches and the strategic deployment of known approaches for the formation of complex heterocycles compounds continue to drive the field of synthetic organic chemistry. Organic chemists have been involved in many efforts to prepare these heterocycles by developing new and efficient synthetic transformations.

Nickel was the first metal catalyst to be used to perform the *N*-alkylation of amines with other amines being the source of the electrophile. This catalytic ability was found when Raney nickel was used as catalyst to perform the reduction of nitrile derivatives. The NO_2 compounds act as important building blocks in the formation of nitrogen-containing heterocyclic compounds because of their high chemical reactivity.

In this book, Raney nickel catalysts and reactions catalyzed by them are explained from a synthetic heterocyclic chemistry point of view.

Index

Page numbers followed by "*f*" and "*t*" indicate, figures and tables respectively.

A

Acid
  assisted process, 108
  catalyzed cyclization, 146
Adams platinum oxide, 109
Albertson's synthesis, 88
Alkoxy-substituted thiophenes, 171
Amide, 105
Amine, 52
Aminoisoxazole derivatives, 149
Appel reaction, 184
Aspidofractine, 188
Azabornane skeleton, 7
Aziridinium salt, 190
Azo compounds, 107

B

Barium hydroxide, 173
Batcho–Leimgruber indole synthesis, 195
Baylis–Hillman acetates, 88
Baylis–Hillman-like reaction, 43
Benzoxazine substrate, 62
Benzoylthioureas, 28
Bischler–Napieralski cyclization, 188
Bischler–Napieralski reaction, 56

C

CDC reaction, 50
Cinchona alkaloid-based catalysts, 56
Clozapine, 71
Comprehensive medicinal chemistry (CMC) record, 159
Cu-catalyzed arylation, 50
Cyanide compounds
  five-membered heterocycles, 82
  synthesis, 82
Cyano derivative, 90
Cyanohydrin synthesis, 160
Cyclic carbamate, 108
Cyclohexenone, 52

D

De novo asymmetric overall synthesis, 192
Desalination process, 164
Dess-Martin oxidation, 188, 189
Dibenzyl azidocarboxylate, 190
Diels-Alder reactions, 133, 178, 188
Diol tetrahydrofuran derivative, 173
Displacement reactions, 34
Domino reactions, 119
Dutch resolution methods, 82

E

Enantiopure dihydropyrroles, 124
Enantioselective decarboxylative reaction, 5
Enolizable compounds, 5
Enzymatic desymmetrizations, 45
Epoxy acid, 201
E-propenyl ether, 164
Ester, 197

F

Fischer indole synthesis, 66
Fumarate-derived nitroalkenes, 134

G

GABA receptor antagonists, 6
Garner's aldehyde, 121, 122, 151
Gibbs free energy rates, 164
Glycosidase inhibitory activity, 133
Grubbs II catalyst, 172

H

Halovinylnitroso compounds, 142
Heck cyclization, 193
Henry reaction, 45
Heterocycles, 43, 81, 119
Heterocyclic compounds, 159
Hetero-Diels-Alder reaction, 142
Heterogeneous catalysts, 43, 119
Heterogeneous nickel catalyst, 1
Hillman reaction, 193

213

Hosomi-Sakurai allylation, 146
2-hydrazino alcohols, 174
Hydrogenation, 162
Hydroxy ester, 178

I

Imidazopyridine-substituted benzimidazole, 33
Indole, 66
Intramolecular alkylation, 138
Isatin, 94
Isocyano-bonding quaternary carbons, 56
Isoquinoline, 93
Isoxazoline product, 141

K

Klebsiella pneumoniae, 176
Kopsinine indole alkaloids, 188, 189

L

Lactam, 53
Lactol, 135
Lewis acids, 145, 192
    promoter, 10
Liquid-phase reactions, 119
Lithio-nitroso-pyrrolidine, 180
Long-term clozapine therapy, 71

M

Mannich reactions, 34
Meerwein's reagent, 164
14-membered cyclopeptides, 151
Metal-catalyzed system, 6, 7
Methylenation reaction, 164
Methylenedioxy group, 70
Michael, 184
    reaction, 56, 93
Mitsunobu reaction, 164
Mosher ester derivative, 45
Multigram scale, 66

N

N-acyl metal enolates, 120
N-containing heterocycles, 81, 195
N-containing heterocyclic compounds, 1, 43
Nectrisine, 102

Nickel, 81
Nitroalkenes, 43
Nitrocyclopentene derivatives, 139
Nitro group reduction, 10, 14, 18, 20, 33
Nitroso acetals
    hydrogenolysis, 131
N-methyl derivative, 103
Norepinephrine reuptake inhibitors, 17
N-substituted indole derivatives, 15
Nucleophilic aromatic substitution, 28
    reaction, 31, 62

O

Ofloxacin, 66
Olefin hydrogenation, 111
Organocatalytic amination reaction, 174
Oxazolidinone, 97, 146
Oxazolone, 85
Oxidative coupling reaction, 50
Oxime compounds, 101, 148
Oxime compounds
    synthesis from, 43
Ozonolysis, 173

P

Peterson-type olefination, 129
    procedure, 129
Pictet-Spengler condensation, 53
Pictet-Spengler cyclization, 105
Pictet-Spengler reaction, 54, 187
Pinnick oxidation, 184
Piperidines, 177
Pomeranz-Fritsch reaction, 180
Pummerer rearrangement-cyclization, 164
Pyranone, 150
Pyrazole ring, 55
Pyrazolone, 85
Pyrazolopyrimidines, 96
Pyrimidoindolone heterocycle, 94
Pyrrolidines, 177
Pyrrolinone, 160
Pyrrolizidine-derived products, 129
Pyrrolizidinone derivatives, 124
Pyrrolobenzodiazepine system, 69
Pyrrolopyrimidone, 162

Q

Quinolizidine, 180

azabicyclic skeleton, 180

R
Raney-Ni catalyst, 93, 131
Raney nickel reduction, 84
Reaction mechanism, 119
Ring-closing metathesis reaction, 146
Robinson annulation reaction, 148

S
Sandmeyer's technique, 15
Schiff base-type condensation reaction, 52
Schmidt reaction, 192
Secondary reaction, 109
Second-resolution method, 62
Seven-membered heterocycles
  synthesis, 97, 200
Six-membered heterocycles, 88, 142
  synthesis, 105, 142
Six-membered N-heterocycles
  synthesis, 177
Six-membered N,N-heterocycles, 193, 195
Six-membered O-heterocycles, 195
Six-step reaction, 61
Spiro-fused diazepane, 99
Stereoselective hydrogenation, 179
Stereoselective reduction, 179
Strecker reaction, 98
Streptomyces melanovinaceus, 176
Sulfonamide, 94
Swern oxidation, 105
Synthetic pathway, 85
Synthetic processes, 121

T
Tertiary amides, 84

Tetraazapyrenes, 55
Tetracyclic amine, 200
Tetraethylammonium chloride
  (TEAC), 19
Thin layer chromatography, 195
Three-step reaction sequence, 188
Toluenesulfonic acid, 94
Tricyclic compounds, 136
Tricyclic system, 200
Triethylsilyl ether, 123
Trifluoroacetic acid, 69

U
Ullmann condensation reaction, 71

V
Vilsmeier-Haack (V-H) reaction, 69
Vilsmeier procedure, 164
Vinylogous urethane, 162

W
Wittig
  cyclization, 176
  homologation, 45, 178
  olefination, 172
  reaction, 146, 197

X
X-ray crystallography, 90, 99

Z
Zinc borohydride, 173

Printed in the United States
by Baker & Taylor Publisher Services